CLIMATE CHANGE AND COTTON PRODUCTION IN MODERN FARMING SYSTEMS

CLIMATE CHANGE AND COTTON PRODUCTION IN MODERN FARMING SYSTEMS

ICAC Review Articles on Cotton Production Research No. 6

M.P. Bange,[1] J.T. Baker,[2] P.J. Bauer,[3] K.J. Broughton,[1] G.A. Constable,[1] Q. Luo,[4] D.M. Oosterhuis,[5] Y. Osanai,[6] P. Payton,[7] D.T. Tissue,[6] K.R. Reddy[8] and B.K. Singh[6]

[1]*Commonwealth Scientific and Industrial Research Organization, Agriculture, Narrabri, New South Wales, Australia;* [2]*USDA-ARS, Big Spring, Texas, USA;* [3]*USDA-ARS, Florence, South Carolina, USA;* [4]*University of Technology Sydney, New South Wales, Australia;* [5]*Department of Crop, Soil, and Environmental Science, University of Arkansas, Fayetteville, Arkansas, USA;* [6]*Hawkesbury Institute for the Environment, Western Sydney University, Penrith, New South Wales, Australia;* [7]*USDA-ARS, Lubbock, Texas, USA;* [8]*Department of Plant and Soil Sciences, Mississippi State University, Mississippi, USA*

www.cabi.org

CABI is a trading name of CAB International

CABI
Nosworthy Way
Wallingford
Oxfordshire OX10 8DE
UK

Tel: +44 (0)1491 832111
Fax: +44 (0)1491 833508
E-mail: info@cabi.org
Website: www.cabi.org

CABI
745 Atlantic Avenue
8th Floor
Boston, MA 02111
USA

T: +1 (617) 682 9015
E-mail: cabi-nao@cabi.org

A catalogue record for this book is available from the British Library, London, UK.

Library of Congress Cataloging-in-Publication Data

Names: Bange, M. P. (Michael P.), author.
Title: Climate change and cotton production in modern farming systems /
    Bange, M.P., Baker [and eleven others].
Description: Boston, MA : CAB International, 2016. | Includes bibliographical
    references and index. | Description based on print version record and CIP
    data provided by publisher; resource not viewed.
Identifiers: LCCN 2015048464 (print) | LCCN 2015045124 (ebook) | ISBN
    9781780648910 (ePDF) | ISBN 9781780648927 (ePub) | ISBN 9781780648903 (pbk
    : alk. paper)
Subjects: LCSH: Cotton--Climatic factors. | Crops and climate. | Sustainable
    agriculture. | Agricultural systems.
Classification: LCC S600.5 (print) | LCC S600.5 .C53 2016 (ebook) | DDC
    630.2/515--dc23
LC record available at http://lccn.loc.gov/2015048464

ISBN-13: 978 1 78064 890 3

Commissioning editor: David Hemming
Editorial assistant: Emma McCann
Production editor: Shankari Wilford

Typeset by SPi, Pondicherry, India
Printed and bound in the UK by CPI Group (UK) Ltd, Croydon, CR0 4YY

CONTENTS

List of Abbreviations vii

Summary ix

I   INTRODUCTION 1

II CLIMATE CHANGE IMPACTS ON MAJOR COTTON
    PRODUCTION REGIONS 1

III CLIMATE CHANGE IMPACTS ON COTTON GROWTH
    AND PRODUCTION 2
    3.1 Impacts on Cotton Physiology, Growth, Yield and Quality 3
         3.1.1 Technologies used in climate change investigations in cotton 3
         3.1.2 Effect of elevated CO_2 concentration 6
         3.1.3 Effect of elevated temperature 11
         3.1.4 Effect of vapour pressure deficit 15
         3.1.5 Effect of drought 15
         3.1.6 Effect of rainfall intensity (flooding/waterlogging) 17
         3.1.7 Interactive effects of climate change 18
    3.2 Climate Change Impacts on Pests and Diseases 19
    3.3 Climate Change Impacts on Soils 20
         3.3.1 Effect of elevated CO_2 concentration 20
         3.3.2 Effect of elevated temperature 21
         3.3.3 Effect of drought 22
         3.3.4 Effect of rainfall intensity (flooding/waterlogging) 22

IV MANAGEMENT APPROACHES TO ADAPT TO IMPACTS
    OF CLIMATE CHANGE 23
    4.1 Cultivar Change 24
    4.2 Season Length and Planting Date 25
    4.3 Pest Management 26
    4.4 Water and Irrigation Management 27
    4.5 Management of Cotton Crops with Plant Growth Regulators 29
    4.6 Crop Diversification with Crop Rotations and Cover Crops 30
    4.7 Utilizing Seasonal Climate Forecasts 31
    4.8 Optimizing Efficiency of Resource Inputs 32
         4.8.1 Crop nitrogen use 32
         4.8.2 Crop water use 33
    4.9 Soil Management 34

V   ROLE OF RESEARCH IN MODERN COTTON
    SYSTEMS ADAPTING TO CLIMATE CHANGE 35
    5.1 Genetic Improvement and Cotton Physiology 36
    5.2 Soil Management 38
    5.3 Cotton System Management 39
         5.3.1 Climate information and use 39
         5.3.2 Policy and industry considerations 39
         5.3.3 Crop management 40

VI CONCLUSION 42

REFERENCES 45

LIST OF ABBREVIATIONS

1-MCP	1-methylcyclopropene
ABA	abscisic acid
ACC	1-aminocyclopropane-1-carboxylate
Adh	alcohol dehydrogenase
AgMIP	Agricultural Model Intercomparison and Improvement Project
anammox	anaerobic ammonium oxidation
AOA	ammonia oxidizing archaea
AOB	ammonia oxidizing bacteria
ATP	adenosine triphosphate
AVG	aminoethoxyvinylglycine
BAP	6-benzylaminopurine
Bt	*Bacillus thuringiensis*
C_i	internal CO_2 concentration
CETA	canopy evapotranspiration and assimilation
CO_2	carbon dioxide
$[CO_2]$	CO_2 concentration
$[eCO_2]$	elevated CO_2 concentration
ENSO	El Niño-Southern Oscillation
FACE	Free Air CO_2 Enrichment
GA_3	gibberellic acid
g_s	stomatal conductance
G×E×M	genetic × environment × management
GHG	greenhouse gas
GSL	growing season length
HEAPS	HElicoverpa Armigera and Punctigera Simulation model
HFI	Horizontal Flowering Index
IAA	auxin-indole-3-acetic acid
IFM	Integrated Fibre Management
IPM	Integrated Pest Management
LWP	leaf water potential
NDVI	Normalized Difference Vegetation Index
NOB	nitrite-oxidizing bacteria
NUE	nutrient use efficiency
OTC	open top chambers
Pdc	pyruvate decarboxylase
PGR	plant growth regulators
PUT	putrescine
QTL	quantitative trait loci
R	rainfall
R_D	dark respiration
Rubisco	ribulose-1,5-bisphosphate carboxylase-oxygenase
RuBP	ribulose-1,5-bisphosphate
RUE	radiation use efficiency
SOI	Southern Oscillation Index
SPAR	Soil-Plant-Atmosphere-Research
SPD	spermidine
SPM	spermine
T	temperature
VFI	Vertical Flowering Index
VPD	vapour pressure deficit
WUE	water use efficiency
WUE_I	instantaneous water use efficiency

SUMMARY

Worldwide, cotton is already broadly adapted to growing in temperate, subtropical and tropical environments, but growth may be challenged by future climate change. Production may be directly affected by changes in crop photosynthesis and water use due to rising CO_2 and changes in regional temperature patterns. Indirect effects of climate change will likely result from a range of government regulations aimed at climate change mitigation. These impacts will also occur in light of other pressures that will be placed on cotton production systems, such as reductions in land and water availability, rising costs of production and a decline in trade as a result of competition from other commodities and man-made fibre.

The essence of this review is to:

1. Summarize the impacts and challenges that climate change will have on cotton production in different regions across the world.
2. Compile and summarize climate change impacts on cotton growth and production.
3. Document research and management practices that may help with adaptation relevant to modern cotton farming systems.
4. Outline research approaches to address climate change.

While there is certainty that future climate change will impact cotton production systems there will be opportunities to adapt. This review begins to provide details for the formation of robust frameworks to evaluate the impact of projected climatic changes, highlight the risks and opportunities with adaptation and detail the approaches for investment in research. Major matters that were identified and discussed in the review were:

- Climate change will have both positive and negative effects on cotton. Increased CO_2 may increase yield in well-watered crops, and higher temperatures will extend the length of growing season (especially in current short-season areas). However, higher temperatures also have the potential to cause significant fruit loss, lower yields and alter fibre quality, and reduced water use efficiencies. Extreme weather events such as droughts, heatwaves and flooding also pose significant risks to improvements in cotton productivity.
- Research into integrated effects of climate change (temperature, humidity, CO_2 and water stress) on cotton growth, yield and quality will require further investment. This includes the development of cultivars tolerant to abiotic stress (especially for more frequent hot, water-deficit and waterlogged situations). Some consideration or allowance will be needed in these studies for both cotton cultivars and insect pests that have been naturally selected in rising CO_2 environments.
- Although cotton is already well adapted to hot climates, continued breeding by conventional means as well as applying biotechnology tools and traits will develop cultivars with improved water use efficiency and heat tolerance. Investment in whole-plant and crop physiology will be important to develop robust understanding of the physiological determinants of cotton crop growth and development. Undertaking this research with the involvement of agronomic researchers, extension specialists, crop managers and growers is vital so that achievements can be recognized in the field as quickly as possible.
- The potential for declining availability of water resources under climate change will increase competition for these resources between irrigated cotton production, other crops and environmental uses. These issues emphasize the need for continual improvement in whole farm and crop water use efficiencies and the need for clear information on water availability.
- There will be a need to improve cotton farm resilience by maintaining and increasing cotton profitability through practices that increase both yield and fibre quality, while improving efficiency of resource use (especially energy, water and nitrogen).
- Region-specific effects will need to be assessed thoroughly so that cotton growers can assess likely impacts at the business level. Also, as cotton is a global commodity it will be vital for cotton industries to understand global changes in cotton markets as part of their overall adaptation strategy.
- Simulation models will play a vital role in assessing impacts and adaptation options for future climate change; however, they will require investment in development and their validation for climate change issues. As new forecasted future climate change scenarios are developed they will also need to be used to update and quantify impacts and re-evaluate adaptation options. Crop biophysical modelling should be appropriately linked to economic whole farm/catchment scale modelling efforts. Similar considerations need to be given to cotton decision support tools that

utilize day-degree functions. It is possible that many systems do not accommodate future predicted extremes associated with climate change (e.g. heatwaves slowing crop development).
- Implementation of whole farm designs that build system resilience through diversity in crops, while increasing soil fertility and protection from erosion through the inclusion of rotation and cover crops will also need further attention.

The review acknowledges that most approaches discussed throughout the review are decidedly production focused, and recognizes that there are other significant efforts to combat 'a changing climate' from other perspectives and scales; policy and catchment scale efforts are some examples. Ultimately, it is a multi-faceted systems-based approach that combines all elements mentioned in the review as well as others that provide the best insurance to harness the change that is occurring, and best allow cotton industries worldwide to adapt. Given that there will be no single solution for all of the challenges raised by climate change and variability, the best adaptation strategy for industry will be to develop more resilient systems. Early implementation of adaptation strategies, particularly in regard to enhancing resilience, has the potential to significantly reduce the negative impacts of climate change now and into the future.

I INTRODUCTION

Cotton is a natural fibre produced by four different species of *Gossypium*. Approximately 95% of the cotton is produced by the *G. hirsutum* L. species; therefore, this review will concentrate primarily on that species, with a few exceptions. Cotton is used every day in the form of clothing made from cotton fibre and products made from cotton-seed oil. Cotton is the most widely produced natural fibre in the world, but there is increasing competition from man-made fibres. Cotton seed is a by-product of the more valuable cotton fibre and is a valued raw material for food oils for human consumption and high protein feed for livestock.

Cotton is a perennial shrub with an indeterminate growth habit and although it grows naturally to 3.5 m in the tropics, it is grown commercially as an annual crop. Wild ancestors of cotton are found in arid regions, often with high daytime temperatures and cool nights, and are naturally adapted to surviving long periods of hot dry weather. Modern cultivars have inherited these attributes, making the cotton crop well adapted to the intermittent water supply that occurs with rainfed (dryland) and irrigated production (Hearn, 1980). Compared with other field crops, however, its growth and development are complex consequences of the indeterminate habit. Vegetative and reproductive growth occurs simultaneously, sometimes making interpretation of the crop's response to climate and management difficult.

Worldwide, cotton is already broadly adapted to growing in temperate, subtropical and tropical environments, but growth may be challenged by future climate change. Production may be directly affected by changes in crop photosynthesis and water use due to rising $[CO_2]$ and changes in regional temperature patterns (Reddy *et al.*, 2000; Oosterhuis, 2013). Indirect effects of climate change will likely result from a range of government regulations aimed at climate change mitigation. These impacts will also occur in light of other pressures that will be placed on cotton production systems, such as reductions in land and water availability, rising costs of production and a decline in trade as a result of competition from other commodities and man-made fibre.

To meet these challenges and opportunities, sustainable cotton production will need to adopt practices, in combination, that will: (i) increase and/or maintain high yield and fibre quality; (ii) improve a range of production efficiencies (water, nutrition and energy); (iii) seek to improve returns for lint and seed; and/or (iv) consider other cropping options as alternatives. Crop management and plant breeding options include: (i) high yielding/high quality stress-tolerant cultivars; (ii) optimizing water; (iii) manipulating crop growth and maturity; (iv) varying planting time; (v) optimizing soil and health for crop nutrition; and (vi) maintaining diligent monitoring practices for weeds, pests and diseases to enable responsive management.

The essence of this review is to:

1. Summarize the impacts and challenges that climate change will have on cotton production in different regions across the world.
2. Compile and summarize climate change impacts on cotton growth and production.
3. Document research and management practices that may help with adaptation relevant to modern cotton farming systems.
4. Outline research approaches to address climate change.

II CLIMATE CHANGE IMPACTS ON MAJOR COTTON PRODUCTION REGIONS

While there is growing confidence in global scale observations and predictions of climate change, it is still difficult to determine precisely how spatial variation of climate change will translate into impacts at regional scales especially in production agriculture. Nonetheless, there are some general principles about the variability of climate change between locations, and current climate models provide some indication of locations where impacts are likely to be most severe.

There have been substantial increases in atmospheric CO_2 concentration ($[CO_2]$) since the beginning of the industrial age. The natural atmospheric $[CO_2]$ during the past 800,000 years ranged between 170 and 300 $\mu mol\ mol^{-1}$ (CSIRO, Bureau of Meteorology, 2012). Atmospheric $[CO_2]$ has increased in the past 200 years from a pre-industrial concentration of about 280 $\mu mol\ mol^{-1}$ to almost 400 $\mu mol\ mol^{-1}$ in 2014 (IPCC, 2013; Tans and Keeling, 2015), with projections for further increases in the future. The rate at which atmospheric $[CO_2]$ is rising is also increasing: global atmospheric $[CO_2]$ increased from 2009 to 2011 at a rate of 2 $\mu mol\ mol^{-1}\ year^{-1}$ (CSIRO, Bureau of Meteorology, 2012).

It is projected that atmospheric $[CO_2]$ will be around 450 $\mu mol\ mol^{-1}$ for the period centred on 2030. There is little difference in projected atmospheric $[CO_2]$ in the near future (i.e. 2030) among

© CSIRO 2016. *Climate Change and Cotton Production in Modern Farming Systems*
(eds M.P. Bange *et al.*)

emission scenarios. However, emission scenarios of projected atmospheric $[CO_2]$ vary considerably with time. For example, atmospheric $[CO_2]$ in 2100 is projected to increase from 500 to 900 μmol mol^{-1} across greenhouse gas (GHG) scenarios (Nakicenovic and Swart, 2000). These scenarios involve assumptions about demographic, economic and technological factors likely to influence future economic development and GHG emissions. Scenarios depend on factors such as rates of population increase, global economic growth and humanity's relative success or failure at slowing emissions from the burning of coal, oil and gas (Braganza and Church, 2011).

III CLIMATE CHANGE IMPACTS ON COTTON GROWTH AND PRODUCTION

Elevated $[CO_2]$ ($[eCO_2]$)-induced climate change could affect cotton production practices and change the historic location of cotton production around the world. Table 1 summarizes some current research efforts, climate indicators and potential results in major cotton-producing

Table 1. Summary of changes in climate and impact indicators for cotton producing regions throughout the world.

Countries	Climate change	Impact indicators	References investigating climate change impacts on cotton systems
China	T 4.5°C by 2080 R increase	Cotton GS Peaking growing period The proportion of opening bolls Fibre quality	China (2004)
China	T increase R increase or decrease (depending on emission scenarios, time periods and locations)	Cotton lint yield Cotton WUE	Yang *et al.* (2014)
India	T increase 2~4°C by 2050 Little change in monsoon R Extreme climate events (i.e. drought, flood) increase	Cotton productivity Fibre quality	Kranthi (2009)
USA	T increase 2.5~5°C by 2080s in south-east Drier conditions in south-west More droughts and floods in west	Cotton yield	NCAR (2001, 2004); Doherty *et al.* (2003)
Pakistan	T increase R decrease	Heat stress Irrigation water requirement	Pakistan (2003)
Uzbekistan	T increase R decrease over central Asia	Cotton crop losses	Uzbekistan (2008)
Brazil	T increase 0.4~1.8°C in 2020; 1~7.5°C in 2080 R reduces −40~−20% and increases 5~10% in 2080	Cotton lint yields	UNFCCC (2008)
West and Central Africa	T increase 3.3°C (end of this century) in West Africa R uncertain An increase in the frequency and intensity of extreme climate events	Cotton production area Cotton distribution	Toulmin (2009)
Turkey	T increase 2~3°C R decrease	Surface water for irrigation	Turkey (2007)
Australia	T increase 0.4~2°C by 2030, 1~6°C by 2070 R change −20~5% in 2030 and −60~10% in 2070	Frequency of heat and cold stress	Bange *et al.* (2010b)
Australia	T increase	Cotton phenology Frequency of heat and cold stress	Luo *et al.* (2014)
Australia	T increase Cotton GS increase, R changes (increases and decreases)	Cotton lint yield Cotton WUE	Luo *et al.* (2015); Williams *et al.* (2015)
Australia	T increase	Whole-farm profitability	Rodriguez *et al.* (2014)
Australia	T increase R increase at harvest	Cotton fibre quality (micronaire and grade)	Luo *et al.*, unpublished

T, temperature; R, rainfall; GS, growing season length; WUE, water use efficiency.

regions. Significant changes in day and night temperatures, altered weather patterns (season length, precipitation patterns, in-season temperature fluctuations), altered evaporative demand, restricted water availability from limited runoff and replenishment of groundwater, changes in pest (insect, weed and disease) species due to shifting temperature regimes and possible changes in geographical regions suitable for cotton (loss of hectarage in some areas and addition of hectarage in other areas) will have direct effects on cotton production practices and profitability. Compounded with the direct effect of atmospheric $[CO_2]$, the resulting impact of climate change on cotton production is difficult to predict. While understanding of cotton's response to these challenges may be partially ascertained from knowledge about cotton grown in currently variable climates in regions throughout the world, we know very little about the interactive effects of these climate drivers on cotton productivity. The current understanding of cotton's growth, development and resource use efficiencies in response to these variables and their interactions is presented here.

3.1. Impacts on Cotton Physiology, Growth, Yield and Quality

Upland (*G. hirsutum* L.) and Pima (*G. barbadense* L.) cotton are indeterminate perennial crops that are cultivated as annuals. As such, their ability to flower over an extended period of time might buffer their reproductive response to climate change. However, their growth, development and performance will still be impacted by the changing environmental landscape. In this section, we will explore the literature describing cotton response to these changing climatic parameters.

3.1.1. Technologies used in climate change investigations in cotton

Several experimental technologies have been used to study the effects of climate change on cotton growth, development and crop productivity, including controlled environmental facilities, Soil-Plant-Atmosphere-Research (SPAR) units, Canopy Evapotranspiration and Assimilation (CETA) chambers, Open Top Chambers (OTC) and FACE (Free Air CO_2 Enrichment) facilities. Experiments conducted in these facilities have provided a large body of useful information on cotton response to climate change. Here, we discuss the relative advantages and limitations of each of these experimental facilities.

Controlled environments

In controlled environment facilities, practical advantages include precise control of $[CO_2]$, humidity, temperature and light, as well as regulating water and nutrition. In general, chamber experiments are valuable for identifying mechanisms of crop response at the molecular, biochemical and physiological scales, but are limited for estimating crop yield (Long *et al.*, 2006) and maturity under field conditions. One concern is that plants grown in pots for long periods of time may have restricted root growth, which in turn can negatively influence photosynthetic capacity, shoot growth and harvestable yield potential and thus reduce the response to CO_2 stimulation that might be expected in unrestricted root growth conditions in the field (Arp, 1991; Ainsworth and McGrath, 2010). Restricted rooting space in pots may also limit nutrient and water availability, thereby reducing the CO_2 response. Finally, single plants are often grown in pots, which do not simulate field conditions where plants are grown in higher density, thereby generating plant competition for water, nutrients and light. In that instance, pot-grown plants may show greater positive response to $[CO_2]$ than in the field. Most of these challenges can be overcome if these limitations are recognized and if the pots are large, nutrients are abundant and plants are harvested before root restrictions are encountered.

There are also some concerns that chamber size, light levels and quality, solar radiation energy and forced air circulation in controlled environments alter plant growth, which compromises the ability to measure accurately crop responses to $[CO_2]$ compared to field-grown plants. The physical size of chambers may sometimes limit the capacity to allow investigators to follow crops to maturity (McLeod and Long, 1999) as plants simply may become too large.

Soil-plant-atmosphere-research

SPAR units were built in the late 1970s at Starkville, Mississippi, USA and studies on cotton growth and development began in 1988 and continue today to a certain extent (Phene *et al.*, 1978;

Reddy, K.R. *et al.*, 1996, 2001). Later, similar facilities were added at the USDA-ARS facilities at Beltsville, Maryland, USA (Fleisher *et al.*, 2009). The primary advantage of SPAR chambers was that they were used outdoors and allowed plants to grow in more natural light, but also had accurate temperature and $[CO_2]$ control. Each SPAR unit consisted of a steel soil bin (100 cm deep, 50 cm wide and 200 cm long) to allow more expansive root growth and a Plexiglas chamber (2.5 m tall and 1.5 m wide, but with 1 m^2 area for canopy development by placing variable density shade cloths around the canopy) to accommodate aerial plant parts, a heating and cooling system, and an environmental monitoring and control system. Therefore, experiments were easily repeatable and the ability to measure and control environmental variables minimized many of the co-varying and confounding factors that occur in the field (Reddy and Hodges, 2000; Reddy *et al.*, 2001, and references cited therein). In addition, the system was capable of measuring whole-canopy gas exchange, while simultaneously controlling $[CO_2]$. Measurements of CO_2 and H_2O fluxes are important for understanding the impacts of the environment on crop productivity. Both single-leaf and whole-canopy gas exchange provide a highly sensitive measure of the impact of environmental stress on crop performance. However, whole canopy net assimilation is more highly correlated with crop growth and final yield than leaf-level measurements of net-assimilation (Reddy, K.R. *et al.*, 1995; Baker *et al.*, 2009). The SPAR units were originally designed to investigate plant responses to the environment including $[eCO_2]$ for modelling purposes (Reddy *et al.*, 1997a; Fleisher *et al.*, 2010). Therefore, many short-term experiments were conducted to generate functional algorithms that were used to develop/improve cotton crop models that ultimately were utilized for climate change impact assessment and analyses (Reddy *et al.*, 2002, 2008; Doherty *et al.*, 2003; Liang *et al.*, 2012a, b). However, few full-season climate change experiments were conducted from seed to maturity including measurements of fibre quality at both ambient and $[eCO_2]$ (Reddy *et al.*, 1997a, c, 1999, 2000; Lokhande and Reddy, 2014a, b, 2015a, b).

Canopy evapotranspiration and assimilation

To overcome limitations associated with controlled environment facilities, studies in cotton have used in-field chambers, OTCs and FACE (Kimball *et al.*, 1997) facilities. In the field, crops are rooted in the ground and exposed to natural sunlight and variation in air temperature. To measure plants growing in cotton fields, SPAR units were modified and developed as CETA chambers (Baker *et al.*, 2009, 2014a), which can actively control atmospheric $[CO_2]$. The CETA chambers operate as open gas exchange systems to measure canopy carbon and water exchange of pot-grown and field-grown cotton plants, and have been used in the USA and Australia. CETA chambers can accurately estimate transpiration (E) across different dates and a wide range of canopy leaf area index (LAI; Baker *et al.*, 2009). One disadvantage of the CETA chambers is that internal air temperature may increase by 2 to 5°C compared with outside ambient air because the air is not actively cooled using an air conditioning system. Instead, CETA chambers use a programmable data logger to control a variable speed fan which may limit heat build-up to 0.5°C above ambient air temperature, provided there is sufficient canopy leaf area and soil water to remove latent energy from the system (Baker *et al.*, 2009).

However, placing a chamber over a crop canopy may change some environmental variables that affect canopy gas exchange inside the chamber (Baker *et al.*, 2014b), compared to cotton growing outside the chamber. In high temperature environments, the fans controlling airflow remain at maximum speed for considerable periods, which may affect leaf gas exchange via changes in the boundary layer resistance and also by mechanical stresses (Kimball *et al.*, 1997). There is most likely a high temperature limit at which removal of latent heat can be achieved by air movement alone without addition of a cooling component. The chamber wall material was also reported to reduce photosynthetically active radiation (PAR) by approximately 13% (Baker *et al.*, 2009).

Open top chambers

OTCs were developed to provide control of atmospheric $[CO_2]$ and temperature, while maintaining natural edaphic conditions of the field setting. In the field, plants can be grown in OTCs, where plants are rooted in the ground and exposed to natural light and precipitation which enter through the top of the chamber. OTC walls are transparent plastic walls, which allow substantial natural light penetration and create a wind barrier which makes it easier to control $[CO_2]$. CO_2-enriched air is introduced to the chamber by a blower system, which generates some air movement simulating natural wind. OTCs are generally small in diameter (3–5 m) to maintain good CO_2 control.

Although OTCs eliminate some of the problems associated with glasshouse and SPAR/CETA chambers, OTCs may alter the environmental conditions, including air temperature and relative humidity which are generally higher, light intensity which is usually lower, and light quality which is different (Ainsworth and McGrath, 2010). Wind speed is often lower and less variable than observed in the field (Nakayama *et al.*, 1994). Differences in temperature and humidity tend to be more pronounced under well-watered conditions (Kimball *et al.*, 1997).

Air temperatures may be controlled in OTCs using active cooling or warming with air conditioning systems. If air temperature is not actively controlled, then observed temperatures inside OTCs are typically 0.5–2.5°C warmer than outside. The degree of temperature rise of the foliage depends on transpiration rates, which strongly depend on environmental variables such as air vapour pressure. Ample ventilation and low vapour pressures can cool foliage temperatures inside OTCs.

In summary, environmental controlled facilities, chambers and OTCs are useful for addressing many research questions, but it should be noted that they may potentially reduce light quantity and quality (Kim *et al.*, 2004), increase the ratio of diffuse light to direct sunlight, modify the vapour pressure deficit either through changes in temperature, humidity, or both, alter air flow, and may intercept rainfall. Access to plants by pests and diseases may be restricted, but if they gain access, higher humidity and more shelter may alter the impact on plants. As a result, changes in the environment due to the facility may affect the response of the plant to [eCO$_2$]. For more detail on the response of cotton to [eCO$_2$] in OTCs, see studies by Kimball and Idso (1983), Kimball *et al.* (1984, 1985) and Kimball and Mauney (1993) conducted in a series of experiments in Arizona, USA from 1983 through 1987, under well-watered and soil water deficit conditions.

Free-air CO$_2$ enrichment

OTCs and transparent-walled chambers, despite being partially open to the atmosphere, have often affected important environmental factors, other than those that were intentionally actively managed. To overcome these limitations, the Free-Air CO$_2$ Enrichment (FACE) system was modified from its original development for forest experiments and developed for low stature crops (Long *et al.*, 2006). Large scale FACE experiments allow the exposure of plants to [eCO$_2$] under natural and fully open-air conditions without significantly altering the micrometeorological conditions around a plot of vegetation. FACE plots may encompass up to hundreds of square metres of vegetation, allowing for use of a buffer zone, which minimizes problems of edge effects. The greater size of FACE plots enables investigation of plant responses to [eCO$_2$] from the genomic to the ecological scale.

FACE systems release CO$_2$-enriched air through vertical vent pipes or pure CO$_2$ through horizontal pipes. The gas is released just above the canopy surface on the upwind side of the plot, rather than utilizing confinement structures. Fast-feedback computer controls adjust the position and amount of CO$_2$ released at different points around the plot, based on measurements of wind speed, direction and [CO$_2$] in the centre of the plot. Therefore, there are limited barriers to light, precipitation, wind, or pests (Nakayama *et al.*, 1994).

FACE systems rely on natural wind and diffusion to disperse the CO$_2$ across the experimental area (Nakayama *et al.*, 1994; Ainsworth and Long, 2005) and therefore these systems encounter problems with accurate CO$_2$ enrichment when wind speeds are low. In addition, FACE systems are large, very expensive to build, operate and maintain, and more suited to large and numerous simultaneous experiments requiring significant personnel needs to successfully capitalize experimental expenditures. Therefore, far fewer FACE experiments have been conducted than controlled environment and SPAR studies. The minimum FACE ring diameter for crops is approximately 6 m, which is large compared with smaller OTCs.

There is evidence to suggest that cyclically varying or surging [CO$_2$], as occurs in FACE studies, may underestimate the response of plants to long-term constant [CO$_2$] with the same mean [CO$_2$] (Bunce, 2012). Responses to pulses of CO$_2$ were related to both the extent of the change and the duration of CO$_2$ enrichment (Evans and Hendrey, 1992), but Holtum and Winter (2003) found lower mean rates of net photosynthesis when [CO$_2$] varied compared with constant [eCO$_2$]. Growth and yield may also be affected if FACE plots are not enriched with CO$_2$ during the night (Bunce, 2014), as is most often the case. Bunce (2014) highlights that the reasons for not undertaking enrichment during the night in FACE experiments include an inability to control [CO$_2$] with low wind in some systems, concerns about the disruption of the microclimate at night, and the cost of CO$_2$. In his work on common bean (*Phaseolus vulgaris* L.) he showed that CO$_2$ enrichment at night had important consequences for growth and yield, and that daytime-only CO$_2$ enrichment greatly underestimated yield increases associated with [eCO$_2$]. Studies comparing the effects of day and night time enrichment, versus daytime only enrichment, have not been conducted with cotton.

Important environmental variables such as temperature, light and humidity are not easily controlled at a field scale in FACE experiments (Kimball *et al.*, 1997). However, infrared heating was added to manipulate canopy temperature in soybean, but this has not been used for cotton (Kimball, 2005; Kimball et al., 2012).

Despite some limitations, FACE data are most likely to represent plant response to future [eCO₂] compared to other facilities, and therefore are valuable for validation of models being developed to predict the effects of increasing atmospheric [CO₂] on plants, ecosystems, agricultural productivity and water resources (Nakayama et al., 1994). Cotton and wheat experiments have shown that *relative* growth responses to [eCO₂] were not significantly different between OTCs and FACE, but the *absolute* growth of cotton was 30% greater inside OTCs (Kimball et al., 1997). Thus, for many studies the FACE approach is preferred because both absolute and relative responses to [eCO₂] can be obtained reliably. FACE facilities were located in Maricopa, Arizona, USA and used to investigate cotton production from 1983 to 1991 (Reddy et al., 2000, and many original references cited therein).

3.1.2. Effect of elevated CO_2 concentration

Increased atmospheric CO_2 concentrations have been shown in cotton to increase leaf photosynthetic rates and crop radiation use efficiency (dry matter produced per unit of intercepted radiation by the crop) and reduce transpiration at the leaf level through reduced stomatal conductance, all potentially leading to improvements in growth and yield.

Photosynthesis and respiration

The increase in crop dry weight depends primarily on the balance between photosynthesis and respiration (Constable and Rawson, 1980a). CO_2 is a primary substrate of the photosynthetic reactions and is fundamental to crop carbohydrate production and overall plant metabolism. In C_3 plants, mesophyll cells containing ribulose-1,5-bisphosphate carboxylase-oxygenase (Rubisco) are in direct contact with the intercellular air space that is connected to the atmosphere via stomatal pores in the epidermis. Rubisco ultimately regulates photosynthetic CO_2 fixation, but has a low affinity for CO_2 such that the carboxylation reaction is not saturated at current atmospheric [CO₂] and therefore will respond positively to rising atmospheric [CO₂] (Drake et al., 1997). Rising [CO₂] also competitively inhibits the oxygenation of ribulose-1,5-bisphosphate (RuBP), which improves the efficiency of net carbon gain by decreasing photorespiratory CO_2 loss (Long et al., 2006).

Short-term gains in photosynthesis due to [eCO₂] are often not maintained at the same magnitude over longer durations. Leaves of plants grown under [eCO₂] frequently undergo quantitative changes in key photosynthetic enzymes and chlorophyll (Harley et al., 1992), and may regulate aspects of the photosynthetic apparatus such as binding of Mn and quinone in photosystem II (Drake et al., 1997). Reductions in photosynthesis over the long term may occur due to reduced quantity of Rubisco protein or a decrease in the activity of the enzyme (Sage et al., 1989). Photosynthetic down-regulation is often due to limitations in soil nutrients (Tissue and Oechel, 1987), where limiting nutrients are allocated away from photosynthesis and into other plant functions, such as growth or maintenance of respiratory function (Ainsworth and McGrath, 2010). If plants are substantially sink-limited (i.e. limited growth capacity), then photosynthetic down-regulation will occur, but generally not completely unless the sink limitation is severe, which rarely happens in cotton systems.

Dark respiration (R_D) of a cotton leaf varies with age. At ambient [CO₂], R_D peaks during rapid growth of the leaf to about 5 ng CO_2 cm⁻² s⁻¹ (1.1 µmol CO_2 m⁻² s⁻¹) then declines to about 1.7 ng CO_2 cm⁻² s⁻¹ (0.3 µmol CO_2 m⁻² s⁻¹) for a 50-day-old leaf (Hearn and Constable, 1984). However, there has been uncertainty surrounding the effects of [eCO₂] on leaf respiration. [eCO₂] was previously suggested to directly inhibit R_D (Reuveni and Gale, 1985; Amthor, 1997; Drake et al., 1997); however, re-evaluation of methods used to measure dark respiration of plants grown under [eCO₂] suggest that short-term exposure to [eCO₂] does not directly affect respiration (Davey et al., 1999; Amthor, 2000; Jahnke et al., 2001; Ainsworth and Long, 2005). If [eCO₂] does affect R_D, it does so through indirect means, such as increased carbohydrate production that increases R_D.

Long-term exposure to [eCO₂] may alter rates of respiration indirectly through the stimulation of photosynthesis, and hence carbohydrate and biomass production. For instance, Reddy, V.R. et al. (1995) found that increased canopy respiration rates in cotton plants grown at 700 µmol mol⁻¹ [CO₂] for 70 days were associated with greater accumulation of biomass and faster growth rate. Leakey et al. (2009) reported that for soybean, greater respiratory quotient and leaf carbohydrate status at 550 µmol mol⁻¹ [CO₂] increased rates of leaf-level respiration at [eCO₂]. Therefore, [eCO₂]

6

Table 2. Percentage increases $(([eCO_2] - \text{ambient } [CO_2])/\text{ambient } [CO_2] *100)$ in net leaf or canopy photosynthesis in saturated light in non-stressed cotton.

Reference	Experiment and (location)	Ambient $[CO_2]$ level (μmol mol^{-1})	$[eCO_2]$ level (μmol mol^{-1})	Increase (%)
Mauney et al. (1978)	Glasshouse (Phoenix, AZ, USA)	330	660	15
Wong (1979)	Glasshouse, leaf (Canberra City, Australia)	330	640	49
Radin et al. (1987)	OTC, leaf (Phoenix, AZ, USA)	350	500, 650	58, 76
Wong (1990)	Glasshouse, leaf (Canberra, Australia)	320	640	59
Harley et al. (1992)	Phytotron (Durham, NC, USA)	350	650	25
Thomas et al. (1993)	Growth chamber, leaf (Durham, NC, USA)	350	650	35–50
Hileman et al. (1994)	FACE, leaf (Maricopa, AZ, USA)	370	550	19, 38, 28 June, July, Aug
Hileman et al. (1994)	FACE, canopy (Maricopa, AZ, USA)	370	550	40, 23, 21 June, July, Aug
Idso et al. (1994)	FACE, leaf (Maricopa, AZ, USA)	'ambient'	500	30
Reddy, A.R. et al. (1998)	SPAR, leaf (Starkville, MS, USA)	350	700	71
Kakani et al. (2004)	SPAR, leaf (Starkville, MS, USA)	360	720	32
Zhao et al. (2004)	SPAR, leaf (Starkville, MS, USA)	360	720	38
Reddy and Zhao (2005)	SPAR, pots, canopy (Starkville, MS, USA)	360	720	65
Bunce and Nasyrov (2012)	Growth chamber, leaf (Beltsville, MD, USA)	380	560	16.6, 7.3, 7.3 1st, 2nd, 3rd leaf
Broughton (2015)	Glasshouse, leaf (Richmond, NSW, Australia)	400	640	28–43
Broughton (2015)	Glasshouse, leaf (Richmond, NSW, Australia)	400	640	19
Osanai et al., unpublished	Glasshouse, leaf (Richmond, NSW, Australia)	400	640	78

may increase respiration rates (on a ground area basis) of cotton grown in future, higher-CO_2 environments as a result of enhanced anabolic processes that consume respiratory adenosine triphosphate (ATP) (Watanabe et al., 2014). However, $[eCO_2]$ may reduce respiration on a dry weight basis if $[eCO_2]$ increases plant mass without changes in leaf-level respiration.

Whole plant growth is often accelerated by CO_2-enrichment, and an increase in growth will generally be accompanied by an increase in growth and maintenance respiration. When growth is increased, growth respiration will increase to supply carbon skeletons, ATP and reductant needed for additional biosynthesis (Amthor, 1991).

Net photosynthesis of a cotton leaf is approximately 30 μmol CO_2 m^{-2} s^{-1} for a recently fully expanded leaf (13–15 days after leaf unfolding, when the leaf was 75–90% of maximum area), under well-watered and fertilized conditions at ambient $[CO_2]$, a leaf temperature of 30°C and saturating light (Constable and Rawson, 1980b; Hearn and Constable, 1984). Plant factors, such as leaf age (photosynthesis declines approx. 12 days after unfurling), and environmental factors alter the rate of photosynthesis; e.g. a 50-day-old leaf has about half the maximum rate of photosynthesis (Constable and Rawson, 1980a). Furthermore, the average leaf age in the whole canopy decreases during the season (Wullschleger and Oosterhuis, 1990). Cotton leaf level and canopy level net photosynthesis is significantly increased in $[eCO_2]$ when plants encounter no nutritional, water, or temperature stress (see Table 2).

Stomatal conductance and transpiration

Guard cells sense changes in $[CO_2]$ at the plasma membrane (Knox et al., 2005) and are thought to respond to the intercellular $[CO_2]$ (C_i) rather than $[CO_2]$ at the leaf surface and in the stomatal pore (Mott, 1988). Electrophysiological studies showed that $[eCO_2]$ alters the activity of K$^+$

channels, which are involved in ion and organic solute concentrations that mediate the turgor pressure in the guard cells (Brearley et al., 1997; Hanstein and Felle, 2002). These changes depolarize the membrane potential of the guard cells and cause stomatal closure (Assmann, 1993; Hanstein and Felle, 2002). Therefore, greater depolarization at [eCO$_2$] will result in a reduced stomatal aperture (Travis and Mansfield, 1979; Macrobbie, 1983; Assmann, 1999; Ainsworth and Rogers, 2007; Abbot et al., 2011). It is expected that guard cell signalling is organized as a network although the signal transduction pathways that function upstream of the ion channel activities are not well known.

One of the most consistent responses of plants to [eCO$_2$] is a decrease in stomatal conductance (g$_s$). Averaged across all plant species grown at [eCO$_2$] in FACE experiments, g$_s$ was reduced by 22% (Ainsworth and Rogers, 2007), although the response of different types of plants (e.g. trees, shrubs, C$_3$ and C$_4$) varied. However, in FACE experiments, the decrease in g$_s$ at [eCO$_2$] does not appear to be caused by a significant change in stomatal density (Estiarte et al., 1994; Reid et al., 2003). Therefore, it is likely that changes in stomatal aperture, rather than density, determine the response of g$_s$ to [eCO$_2$] (Ainsworth and Rogers, 2007). While the sensitivity of guard cells to environmental factors does not appear to acclimate with growth at [eCO$_2$], the magnitude of the effect of higher [CO$_2$] on g$_s$ varies considerably with environmental factors. There is generally a smaller effect of [eCO$_2$] on g$_s$ during dry periods (Ainsworth and Rogers, 2007).

Research in cotton at a leaf level has demonstrated reductions in stomatal conductance and transpiration (see Table 3) with [eCO$_2$]. Responses of transpiration to [eCO$_2$] are varied, particularly between SPAR and FACE experiments. In a SPAR study, Reddy, K.R. et al. (1995) showed that whole canopies of cotton plants grown in high (700–900 µmol mol^{-1}) [CO$_2$] environments transpired less water than plants grown in low (350 µmol mol^{-1}) [CO$_2$] conditions, under optimal conditions (in the absence of water deficit). Similarly, transpiration per unit leaf area was lower at [eCO$_2$] (710 µmol mol^{-1}) compared with plants grown at ambient [CO$_2$] (352 µmol mol^{-1}) for plants grown in an environmental chamber (Phytotron).

Table 3. Reported responses in leaf g$_s$, WUE$_l$ or canopy transpiration for cotton experiments undertaken with ambient and elevated [CO$_2$] in non-stressed cotton expressed as a percentage change (([eCO$_2$] − ambient [CO$_2$])/ambient [CO$_2$] *100).

Reference	Experiment and (location)	Ambient [CO$_2$] level (µmol mol^{-1})	[eCO$_2$] level (µmol mol^{-1})	g$_s$ (% change)	WUE$_l$=A/E (% change)	Canopy transpiration (% change)
Wong (1979)	Glasshouse (Canberra City, Australia)	330	640	–	89	–
Morison and Gifford (1984)	Phytotron (Canberra, Australia)	Ambient	Ambient +340		–	+4.4
Harley et al. (1992)	Phytotron (Durham, NC, USA)	350	650	−32	–	–
Hileman et al. (1994)	FACE Canopy (Maricopa, AZ, USA)	370	550	–	+41, +23, −1 June, July, Aug	–
Samarakoon and Gifford (1995)	Phytotron, pots (Canberra, Australia)	352	710	–	+61, +71 H$_2$O/dry weight	–
Reddy, K.R. et al. (1998)	SPAR, leaf (Starkville, MS, USA)	350	700	−30	–	–
Zhao et al. (2004)	SPAR, leaf (Starkville, MS, USA)	360	720	−34	+55	–
Bunce and Nasyrov (2012)	Growth chamber, leaf (Beltsville, MD, USA)	380	560	−14.9, −19.2, −43.3 1st, 2nd, 3rd leaf	–	–
Broughton (2015)	Glasshouse, leaf (Richmond, NSW, Australia)	400	640	−35	+55	+26
Broughton (2015)	CETA field (Narrabri, Australia)	390	630	−25	–	–
Osanai et al., unpublished	Glasshouse (Richmond, NSW, Australia)	400	640	+30	+72	–

In FACE studies, Bhattacharya *et al.* (1994) demonstrated that CO_2 enrichment decreased stomatal conductance and single-leaf transpiration only towards the end of the season. It has been suggested that CO_2-enrichment should increase water use efficiency (WUE) because [eCO_2] increases biomass and causes partial stomatal closure at the same time, consequently reducing transpiration (Kimball and Idso, 1983; Mauney *et al.*, 1994). Increasing [CO_2] can improve WUE of single, sunlit leaves (Hileman *et al.*, 1994). An increase in [CO_2] from 400 µmol mol^{-1} to 600 µmol mol^{-1} increased instantaneous WUE (WUE_I) by 30–40% (Ko and Piccinni, 2009). In more recent studies by Broughton (2015), WUE_I was increased by 55% in [CO_2] from 400 µmol mol^{-1} to 640 µmol mol^{-1}.

When whole canopy transpiration is assessed, the situation is different (Table 3). Hileman *et al.* (1994) found that canopy transpiration generally was not affected by CO_2 enrichment, except for late in the second season, as the decrease in leaf stomatal conductance was negated by an increase in canopy size. This suggests that cotton crops grown in future, higher CO_2 environments may have increased photosynthetic rates and greater yields, but will require the same amount of water as crops grown under current conditions; however, this study did not account for changes in temperature. Similarly, other FACE experiments found that [eCO_2] (550 µmol mol^{-1}) did not significantly change cotton crop transpiration (Dugas *et al.*, 1994; Hunsaker *et al.*, 1994; Kimball *et al.*, 1994). Studies using water balance evaporation (Hunsaker *et al.*, 1994), sap flow (Dugas *et al.*, 1994) and energy balance (Kimball *et al.*, 1994) methods for measuring transpiration of cotton found that there was no significant difference in canopy evapotranspiration of cotton grown in at [eCO_2] (550 µmol mol^{-1}) compared with ambient [CO_2]. This suggests that [eCO_2] may decrease transpiration at the leaf level, but increased overall plant size and leaf area may not equate to reduced water use at the plant and crop level. However, altered plant response to high frequency CO_2 pulses in FACE studies (Bunce, 2012) may partially explain the differences in canopy transpiration rates of plants grown with [eCO_2]. Therefore, there is still much uncertainty regarding the effect of [eCO_2] on canopy transpiration rates, and thus plant water use, in cotton.

Reddy *et al.* (2002) also noted that [eCO_2] does not noticeably increase transpiration at the canopy level. The decrease in stomatal conductance at the leaf level is offset by the increase in leaf area (Reddy *et al.*, 2000), with the net effect being a negligible difference in canopy transpiration. This suggests that cotton crops grown in future, higher CO_2 environments may have increased photosynthetic rates and greater vegetative growth and yields, but will require at least the same amount of water as crops grown under current conditions; however, this does not account for changes in air temperature and the interactive effects on water use. In situations where water is limited, it is possible that potential increases in transpiration caused by increased leaf growth due to [eCO_2] and elevated air temperature early in the season may result in plants that use more water and subsequently have less water available for reproductive growth; this issue is discussed later in more detail. Reported responses in g_s, WUE_I or canopy transpiration for cotton experiments undertaken with [eCO_2] are summarized in Table 3.

Phenology

The developmental events of cotton plants, such as floral initiation, days to first flower, the rate of mainstem node production (Reddy, K.R. *et al.*, 1995, 1997a, b) and individual leaf area expansion duration (Reddy *et al.*, 1997c) have been reported to be relatively insensitive to increased CO_2 (Reddy *et al.*, 1997c, 2000). Boll maturation period was also not affected by atmospheric [CO_2] (Reddy *et al.*, 1999). However, Osanai *et al.* (unpublished) found that [eCO_2] (640 ppm) accelerated the rate of development, particularly during the early development (i.e. first square and first flower). Production of vegetative branches and number of leaves produced on branches and total fruiting sites produced were mainly controlled by temperature and modulated by carbon supply (Reddy *et al.*, 1997b, 2000).

Growth, yield and fibre quality

It has been estimated that the photosynthetic rate of C_3 and C_4 agricultural crops would increase by 33% and 10%, respectively, with a doubling of the [CO_2] (Kimball, 1983; Cure and Acock, 1986). In addition to enhancing leaf and canopy CO_2 assimilation, CO_2 is also a competitive inhibitor of photorespiration. As a result growth, yield and leaf photosynthetic rates of cotton all respond strongly to CO_2 enrichment (Hileman *et al.*, 1994).

Increasing atmospheric [CO_2] increases carbohydrate pools in cotton (Hendrix *et al.*, 1994). These pools can be used during periods of high metabolic demand, such as heavy fruit set or root growth, to allow such plants to be able to resist metabolically stressful periods better than plants grown in ambient [CO_2]. Cotton grown in CO_2-enriched environments have increased leaf, stem and root carbohydrate content compared with those grown under ambient CO_2 conditions (Hendrix et al., 1994). Doubled atmospheric (720 μmol mol⁻¹) $[CO_2]$ caused plants to produce 40% more leaf, stem and root mass than plants grown at 360 μmol mol⁻¹ $[CO_2]$ (Reddy et al., 1997a). CO_2 enrichment from 370 to 550 μmol mol⁻¹ increased season-long biomass accumulation by 39% under full irrigation (Bhattacharya et al., 1994). Similarly, Mauney et al. (1994) found that increasing $[CO_2]$ to 550 μmol mol⁻¹ in FACE experiments increased biomass by 37%. This was a result of increased crop radiation use efficiency (dry matter per unit of intercepted radiation), which improved on average by 26% from 1.56 to 1.97 g MJ⁻¹ of intercepted radiation over ambient $[CO_2]$ (Pinter et al., 1994b).

Varying CO_2 treatments did not change the ratio of dry matter partitioned to the roots of cotton plants (Reddy, K.R. et al., 1995). However, Prior et al. (1994) stated that increases in atmospheric $[CO_2]$ will enhance absolute plant root growth, with taproots of CO_2-enriched plants displaying greater volume, dry weight, length and tissue density than those grown at ambient $[CO_2]$.

Plants grown in high $[CO_2]$ produced more vegetative branches and more secondary fruiting branches than plants grown in ambient CO_2 environments (Reddy, K.R. et al., 1995). Cotton grown at 720 μmol mol⁻¹ atmospheric $[CO_2]$ had about 40% more squares and bolls than at 360 μmol mol⁻¹ $[CO_2]$ (Reddy et al., 1999), resulting in lint yield increasing by up to 60% in well-watered conditions at moderate temperatures. Harvestable yield was also increased 43% when $[CO_2]$ was increased by 48% in FACE experiments (Mauney et al., 1994). Both increased biomass and yield can be attributed to increased early leaf area (direct effect), and a more profuse flowering and a longer period of fruit growth (indirect effect). Reported responses in parameters for growth and yield for cotton experiments undertaken with $[eCO_2]$ in non-stressed situations are summarized in Table 4.

Table 4. Reported responses in growth and yield for cotton experiments undertaken with ambient and elevated $[CO_2]$ in non-stressed situations expressed as a percentage change $(([eCO_2] - $ ambient $[CO_2])/$ambient $[CO_2] *100)$.

Reference	Experiment and (location)	Ambient $[CO_2]$ level (μmol mol⁻¹)	$[eCO_2]$ level (μmol mol⁻¹)	Biomass (% increase)	Maximum leaf area (% increase)
Mauney et al. (1978)	Glasshouse (Phoenix, AZ, USA)	330	630	110	91
Wong (1979)	Glasshouse (Canberra City, Australia)	330	640	100	60
Morison and Gifford (1984)	Phytotron (Canberra, Australia)	Ambient	Ambient +340	–	9.2 (n.s.)
Delucia et al. (1985)	Phytotron (Durham, NC, USA)	350	675, 1000	72, 115	–
Thomas and Strain (1991)	Growth chamber (Durham, NC, USA)	350	650	46	–
Harley et al. (1992)	Phytotron (Durham, NC, USA)	350	650	30	–
Thomas et al. (1993)	Growth chamber (Durham, NC, USA)	350	650	107	20
Mauney et al. (1994)	FACE (Phoenix, AZ, USA)	370	550	34–37	–
Samarakoon and Gifford (1995)	Phytotron (Canberra, Australia)	352	710	–	70–80
Reddy, A.R. et al. (1998)	SPAR (Starkville, MS, USA)	350	700	30.6	39.5
Zhao et al. (2004)	SPAR (Starkville, MS, USA)	360	720	–	47
Reddy and Zhao (2005)	SPAR (Starkville, MS, USA)	360	720	36.2	18.2

(n.s.), not significantly different

In line with changes in transpiration rates for canopies under [eCO$_2$], Mauney *et al.* (1994) reported increased WUE (kg lint mm^{-1} evapotranspiration) as a function of increasing biomass production rather than a reduction in water use in the FACE experiments. For efficiencies relating to nutrient use in [eCO$_2$], there are reported reductions in tissue nutrient concentration in cotton (Huluka *et al.*, 1994); however, there were no reported differences in nutrient retrieval and utilization (Pinter *et al.*, 1994a) from plants taken from the field in the FACE experiments. Reddy, K.R. *et al.* (1998) reported that cotton plants grown in the SPAR facilities had lower N concentration in leaves, but this had no deleterious effect on lint yields. It is most likely that rapid growth caused by [eCO$_2$] causes growth dilution (less nutrient per unit of biomass), but retrieval and utilization is enhanced by improved root and leaf area.

Elevated [CO$_2$] did not affect fibre parameters, such as micronaire and fibre length, strength and uniformity, that are important to the textile industry in studies by Reddy *et al.* (1999); however, fibre maturity (a component of micronaire) was not measured. Further data are required on effects of fibre quality to expand cotton model predictions with climate change (Yoon *et al.*, 2009). Factors such as changes in temperature regimes and water are most likely to have more significant effects on fibre quality that pertain to both fibre lengthening and fibre maturation (Lokhande and Reddy, 2014a, b).

3.1.3 Effect of elevated temperature

Temperature has two main influences on cotton growth and development. First, it determines rates of morphological development and crop growth (e.g. nodal development, rate of fruit production, photosynthesis and respiration) (Hearn and Constable, 1984). Second, it determines the length of a growing season. There is no clear optimum temperature for cotton because it varies with plant developmental stage, plant organ, and the climate of origin for the cultivar (Burke and Wanjura, 2009). Therefore, the effect of rising temperature may result in both positive and negative effects across cotton-growing regions.

Cotton, in its native state, grows as a perennial shrub in a semi-desert habitat and requires warm temperatures. However, despite originating from hot climates, cotton does not necessarily grow and yield best at very high temperatures. Indeed, a negative correlation has been reported between yield and high temperature during flowering and early boll development (Oosterhuis, 1999; Pettigrew and Oosterhuis, 2013). Although cotton is sensitive to high temperature at all stages of development, it is particularly sensitive during reproductive development, and environmental stress during floral development represents a major limitation to crop development and productivity (Reddy *et al.*, 1992a, b, 1997a; Snider *et al.*, 2010; Oosterhuis and Snider, 2011). Furthermore, the effects on growth from elevated minimum temperatures during the night may be of more importance than during the day and need to be considered (Jagadish *et al.*, 2014). High temperatures can have both direct inhibitory effects on growth and yield and indirect effects due to high evaporative demand with high temperatures generating more intense water stress (Hall, 2001).

Photosynthesis and respiration

Plant tissue temperature plays an important role in both photosynthesis and respiration. Temperature affects all biological activity because it determines the rates of chemical reactions and the activity of enzymes (Knox *et al.*, 2005), including Rubisco. It has been proposed that the regeneration of RuBP, via energy supplied from the electron transport chain, is also a primary limitation to photosynthesis (Stidham *et al.*, 1982; Salvucci and Crafts-Brandner, 2004). At higher leaf temperatures it decreases the relative specificity of Rubisco for CO$_2$ compared with O$_2$ due to decreased solubility of CO$_2$ relative to O$_2$ (Ku and Edwards, 1977) and decreases the affinity of Rubisco for CO$_2$ relative to O$_2$ (Drake *et al.*, 1997; Salvucci and Crafts-Brandner, 2004).

Increases in tissue temperature also increase the rates of both photorespiration (Berry and Bjorkman, 1980) and dark respiration in cotton (Harley *et al.*, 1992). This can ultimately result in lower translocation rates to developing sinks. Hearn and Constable (1984) suggested that maintenance respiration is essentially doubled for each 10°C increase in air temperature affecting tissue. Salvucci and Crafts-Brandner (2004) measured dark respiration and showed that respiration of a cotton leaf substantially increases with air temperature from 28°C to 42.5°C. Night air temperatures above 25°C have been shown to increase the rates of respiration, which results in decreased soluble carbohydrate concentrations in source leaves, increased fruit abscission (Oosterhuis and Snider, 2011) and less plant growth (Arevalo *et al.*, 2008).

The optimal tissue temperature for gross photosynthesis in cotton is approximately 30°C (Reddy, A.R. *et al.*, 1998; Bednarz and van Iersel, 2001; Perry *et al.*, 2013) with an ideal range between 23.5°C and 32°C for metabolic activity and photosynthesis (Burke *et al.*, 1988). A recent study by Conaty *et al.* (2012) demonstrated that a high-yielding Australian cultivar (Sicot 70BRF) also exhibited a similar response. Hence high air temperatures (>35°C) throughout the season may limit photosynthetic potential for plant growth and yield (El-Sharkawy and Hesketh, 1964; Lu *et al.*, 1997; Bibi *et al.*, 2008a). While cotton leaf net photosynthesis exhibits a distinct optimum, cotton canopy photosynthesis is less sensitive to daytime temperatures in well-watered conditions. This is because the canopy itself affects the prevailing atmospheric environment (temperature and humidity) surrounding the canopy (Hearn and Constable, 1984). For example, in hot and dry environments (high vapour pressure deficit (VPD)) air temperatures around a canopy can be cooler than the air away from the canopy because of evapotranspiration. However, canopy temperatures rise when cotton crops are subjected to increasing water deficits and increasing atmospheric VPDs and may result in decreased photosynthesis (Conaty *et al.*, 2014).

Stomatal conductance and transpiration

Transpiration of water from plant leaves cools them due to the large latent heat of vaporization of water relative to the heat capacity (Wiegand and Namken, 1966). During daylight hours, incoming radiant energy is dissipated by transpiration, thereby moderating leaf temperatures; subsequently, higher transpiration rates are often observed at higher air temperatures (Burke and Upchurch, 1989).

The effects of temperature on stomatal aperture which lead to differences in transpiration have been demonstrated in cotton, and has been described as an active process throughout the air temperature range of from 5 to 50°C (Burke and Upchurch, 1989). This response results in large effects on stomatal conductance (g_s) and internal CO_2 concentrations (C_i) of the cotton leaf, with cooler temperatures leading to lower g_s and C_i (Radin and Ackerson, 1981). Changes in stomatal aperture impact C_i of the leaf (Radin *et al.*, 1987), which may lead to changes in photosynthesis and respiration rates as a result of CO_2 substrate availability to Rubisco. Reddy, K.R. *et al.* (1995) also demonstrated that increasing temperature increased the leaf transpiration rate of cotton, and that transpiration rates varied throughout the day in accordance with gas exchange. This study also showed that although increasing temperature increased the rate of leaf transpiration, CO_2 enrichment reduced transpiration at all temperature regimes evaluated.

Adequate availability of water is crucial in the field because heat stress often accompanies drought stress. Some studies have suggested that selection for high yield in irrigated conditions indirectly selected for higher stomatal conductance, resulting in lower leaf temperatures and avoidance of heat stress (Lu *et al.*, 1998). Well-watered plants keep their stomata open at high temperatures, such that evaporative cooling reduces the temperature of the leaves. This process works extremely well in hot, arid regions where there are generally large differences between air and leaf temperatures. Field and glasshouse studies have shown cotton maintained leaf temperatures between 27 and 32°C when there was adequate water available for transpiration cooling at high temperatures (Burke and Upchurch, 1989). However, the efficiency of leaf cooling by transpiration decreases with increasing relative humidity (lower VPD), or when transpiration slows because of water deficit (Salvucci and Crafts-Brandner, 2004). The impacts of humidity on stomatal conductance, transpiration and assimilation are further discussed in the section on vapour pressure deficit.

Phenology

The primary factor affecting cotton development is temperature. As for many crop species, the rate of developmental processes in upland cotton (*G. hirsutum* L.) is strongly responsive to temperature (Hearn and Constable, 1984; Hodges *et al.*, 1993). Cotton is an indeterminate species, so there are less clearly defined phenological events at the whole plant level which can be described directly as a function of temperature. For example there is no clearly defined flowering period and maturity. Beyond first-square (flower bud), there are no phenological events at the whole plant level that can be described solely as a function of temperature. Rather, because cotton is an indeterminate species, the length of the flowering period and timing of crop maturity appears to be governed by a balance between the demand of the developing fruit (boll) load and the plant's capacity to support it. This follows the nutritional hypothesis of Mason (1922) supported by crop-level measurements (Hearn, 1969, 1972, 1994). After the time of first-square, development comprises the ongoing production of mainstem nodes and the extension of the sympodia exerted from these nodes.

At an organ level, development of individual fruit is influenced by temperature and can be described as occurring in two separate periods: the square period (the period between the appearance of a square and anthesis) and the boll period (the period between anthesis and dehiscence). These periods are principally influenced by temperature and by their position on the cotton plant (Constable, 1991).

The pattern of flowering is a product of the interval between flowering at consecutive positions along the sympodial branches (Horizontal Flowering Index – HFI) and the interval between flowering at the first position on consecutive sympodia (Vertical Flowering Index – VFI). The potential rate of both of these processes is governed by temperature at the organ level, but with increasing number of fruit on the plant, the rate of development is reduced and finally ceases. This cessation is presumably due to an inadequate supply of photosynthate (Guinn, 1974, 1985; Mauney et al., 1978), although the involvement of plant hormones cannot be excluded (Guinn and Brummett, 1989).

All developmental events are accelerated with rising temperature. In studies summarized by Hesketh et al. (1972), rates of development increased with a rise in average daily temperature of 15–30°C; similar findings were observed by Hodges et al. (1993). Average daily temperatures above 30°C generally accelerated VFI, HFI, square and boll periods, and emergence to first-square. Similar results were reported for faster appearance of first-square above 30°C by Bange and Milroy (2001) and Reddy et al. (1997a).

In cotton, temperature regulates the rate of development, biomass accumulation, and determines the start and end of the growing season (Baker et al., 1972; Hearn and Constable, 1984). Increased temperature, at the beginning and end of the season, may have a positive effect on yield by extending the duration for cotton growth. Climate change has the potential to raise minimum temperatures and reduce the number of incidences of chilling injury at the start of the season and delay cut-out (last effective flower) (Bange et al., 2010b; Luo et al., 2014). However, an increase in the frequency of days and nights with very high temperatures in excess of an optimal day/night temperature regime may have a negative impact on growth, development and yield (Stockton and Walhood, 1960). The ideal daytime temperature range for cotton is from 20°C to 30°C (Reddy et al., 1991b), although cotton may be grown successfully at temperatures exceeding 40°C in India and Pakistan (Loka et al., 2011), albeit with lower yields.

Growth, yield and fibre quality

The optimum air temperature for stem and leaf growth of cotton is about 30°C (Hodges et al., 1993), and once temperatures reach above 35°C, leaf area declines (Reddy et al., 1992b; Bibi et al., 2010a). Decreased biomass at air temperatures above the optimum range at 30°C/20°C was partly due to higher respiration rates (Reddy et al., 1991a). Shoot biomass of plants at higher air temperatures (40°C/30°C) was severely limited. Higher temperature results in an increase in vegetative branches and a decrease in fruiting branches (Hodges et al., 1993). Three-week-old plants exposed to high air temperatures for 4 days had reduced biomass production, associated with an inhibition of net photosynthesis (Crafts-Brandner and Salvucci, 2004). However, it is sometimes difficult to distinguish the effects of temperature from water deficit stress as high VPDs are often associated with high temperatures (Hearn and Constable, 1984). Roots generally have a lower optimum temperature range for growth than shoots, with optimum air temperatures reported to be 30°C (Arndt, 1945; Pearson et al., 1970). McMichael and Burke (1994) showed that root growth was enhanced when the root tissue temperatures were within or below cotton's thermal kinetic window of 23.5 to 32°C (Burke et al., 1988).

Reproductive development is particularly sensitive to high air temperature both before and after anthesis in cotton (Reddy, K.R. et al., 1996, 2005; Oosterhuis, 2002; Reddy and Zhao, 2005). Boll number and boll size, the basic yield components, are negatively impacted by high temperature and thus final yield has also been shown to be strongly influenced by air temperature in cotton (Wanjura et al., 1969). High, above-average air temperatures during the day can decrease photosynthesis and carbohydrate production (Bibi et al., 2008a), and high night temperatures may increase respiration and further decrease available carbohydrates (Gipson and Joham, 1968b; Loka and Oosterhuis, 2010), resulting in increased fruit abscission (Reddy et al., 1999; Arevalo et al., 2008), decreased seed set, reduced boll size from decreased number of seeds per boll and the number of fibres per seed (Arevalo et al., 2008).

In studies summarized by Hodges et al. (1993), the total number of fruiting sites increased approximately 50% as the air temperature increased from 30°C to 40°C, whereas at temperatures above 35°C abscission of bolls increased sharply with near zero retention of bolls at 40°C. Boll retention decreases significantly under high temperature (Reddy et al., 1991b; Zhao et al., 2005) and is reported to be the most heat sensitive component of cotton yield, with enhanced abortion of squares and young bolls at temperatures above 30°C for both Pima and upland cotton (Reddy *et al.*, 1991b). Most bolls were abscised at high temperatures 3–5 days after anthesis (Reddy *et al.*, 1999) and this has been associated with substantial alterations in the carbohydrate balance of reproductive tissues and vitality of pollen (Reddy *et al.*, 2005). Abscission rates have been negatively correlated with the non-structural carbohydrate content of young bolls (Loka *et al.*, 2011). In field conditions, loss of fruit may cause the crop to grow vegetative (rank) following the period of heat stress. While delaying crop maturity, this compensatory growth response may promote additional fruit growth and reduce negative impacts on final yield; however, this is dependent on remaining season length (light and temperature) that may facilitate growth (Wilson *et al.*, 2003).

High temperature stress is a major factor negatively impacting seed development. Depending upon the duration, timing and severity of the heat stress, fertilization could be limited by poor pollination, decreased pollen germination and limited pollen tube growth. Sensitivity of reproductive organs to heat stress was attributed to the sensitivity of pollen grains to high temperature extremes. A positive correlation exists between anther sterility and maximum temperatures at 15 and 16 days prior to anthesis (Meyer, 1966). However, Barrow (1983) reported that pollen viability and germination were unaffected by pre-treating pollen with temperatures as high as 40°C, whereas penetration of the stigma, style and ovules was negatively impacted at 33°C and above. Snider *et al.* (2011) confirmed that pollen tube growth rate was more sensitive to high temperature than processes occurring during anthesis. The optimal temperature range for cotton pollen germination is between 28 and 37°C , whereas the optimal temperature for pollen tube growth is from 28 to 32°C for a range of *G. hirsutum* cultivars (Burke *et al.*, 2004; Kakani *et al.*, 2005).

A visible consequence of direct tissue damage from severe heat stress is parrot-beaked bolls (Evenson, 1969). This results in small bolls with uneven seed numbers between the locules caused by poor pollination and seed set particularly in one locule. Where substantial numbers of bolls are affected, yield may be reduced. There are no known studies addressing whether the plant compensates for parrot-beak bolls by having other normal bolls grow bigger. Poor fertilization efficiency under high temperature accounts for the decline in seed number observed for cotton exposed to high temperature conditions in both the field (Pettigrew, 2008) and the growth chamber (Snider *et al.*, 2009; Bibi *et al.*, 2010b), and can be a contributing factor affecting boll size (Pettigrew, 2008).

In studies undertaken in the SPAR facilities, boll growth weight was optimum at 30°C/20°C (day/night) temperatures and was reduced at both higher and lower temperatures (Reddy *et al.*, 1991b; Reddy, K.R. *et al.*, 1992a). At lower temperatures, this was associated with low boll growth rates, and at higher temperatures the boll retention was lower and remaining bolls were those produced later in plant development. In further studies undertaken by Reddy *et al.* (1999), boll growth rates were highest at average temperatures around 23°C, although greater boll sizes were measured at 18°C, and this was associated with a longer boll maturation period. Hesketh and Low (1968) showed that boll weight was lower at average temperatures above 24.5°C and was associated with both lower seed weight and smaller boll maturation periods. They also found variation amongst genotypes.

Cotton is a perennial crop and cultivated species are generally photoperiod insensitive, so warmer temperatures at the beginning and end of the season will increase the length of the growing season, provided adequate water and crop nutrition are available. For every extra week of growth period (time between planting and maturity (60% open bolls)), there is the potential to increase lint yield by between 68 and 136 kg ha^{-1} (Bange and Milroy, 2004).

Changes in temperature can have significant effects directly or indirectly (through changes in boll setting patterns) on cotton fibre quality (Bradow and Davidonis, 2000). Fibre length is negatively affected by sustained periods of high temperatures because the period for fibre elongation is reduced, preventing attainment of the genetic potential. Stockton and Walhood (1960) found that as boll temperatures increased above 32°C, fibre length was reduced. In experiments undertaken in the SPAR facilities, Reddy *et al.* (1999) showed that as average air temperatures during the boll maturation period increased from 18°C to 32°C fibre length was reduced. At an average temperature of 26°C, short fibre content (percentage of fibres less than 12.7mm) was lowest and length uniformity (ratio of mean fibre length to the upper half mean length) was highest. They suggested that reductions in fibre length were a result of reductions in the rate of fibre elongation and/or duration of elongation.

The degree of fibre thickening, or fibre maturity associated with secondary fibre cell wall thickening, contributes to differences in micronaire. When comparing fibres of similar perimeter, the thicker the layers of cellulose formed the more mature the fibre, and the higher the micronaire. Since fibre is primarily cellulose, impacts on net crop photosynthesis and carbohydrate production will influence fibre thickening; sustained changes in temperature during the fibre thickening period will lead to differences in micronaire (Bradow and Davidonis, 2010). Many studies demonstrate that micronaire responds directly to temperature (Gipson and Joham, 1968a; Hesketh and Low, 1968; Gipson and Ray, 1970; Wanjura and Barker, 1985; Liakatas *et al.*, 1998; Reddy *et al.*, 1999; Bange *et al.*, 2010a). Most studies observed that micronaire increased with temperature, but Hesketh and Low (1968) and Reddy *et al.* (1999) found a reduction in micronaire with very high temperatures. Reddy *et al.* (1999) observed that micronaire increased with increasing average temperature up to 26°C, but decreased at 32°C and above. Hesketh and Low (1968) observed that micronaire increased up to average temperatures of 30.5°C, but decreased at 33.5°C. At a crop level, shifting the fruiting period within the season and exposing crops to different temperature regimes have demonstrated changes in micronaire (Bange *et al.*, 2008; Bradow and Davidonis, 2010; Braunack *et al.*, 2012).

Large increases in temperature may reduce the number of fruiting branches and the interval between flowering and boll opening, thereby shortening the time to maturity and reducing yield. This may increase final micronaire by limiting the number of late-set bolls that can have lower micronaire. The consequences of higher temperatures on yield and quality are greater if water stress also occurs during these periods.

3.1.4. *Effect of vapour pressure deficit*

Vapour pressure is determined by air temperature and humidity. VPD is the difference between moisture in the air and the amount of moisture the air can hold when it is saturated (Bureau of Meteorology, 2011). Therefore, changes in temperature and relative humidity will affect VPD, which varies throughout the day and across seasons and regions (Pettigrew *et al.*, 1990).

The principal factor affecting transpiration is leaf-to-air vapour pressure difference (Grantz, 1990). As the VPD between leaf and air increases, stomata generally respond by partial closure. Transpiration is affected by changes in stomatal aperture and conductance for water vapour loss, and vapour pressure gradient between the ambient air and sub-stomatal cavity (Hatfield *et al.*, 2011). By mitigating high transpiration that would otherwise be generated by increasing VPD (Rawson *et al.*, 1977), stomatal closure avoids the corresponding decline in plant water potential (Oren *et al.*, 1999), and in most cases, stomatal conductance decreases exponentially with increasing VPD (Farquhar and Sharkey, 1982; Baker *et al.*, 2007). High VPDs result in lower WUEs (Stokes and Howden, 2010). Similar reductions in stomatal conductance have also been found in cotton leaves (Yong *et al.*, 1997; Duursma *et al.*, 2013; Conaty *et al.*, 2014). Duursma *et al.* (2013) showed that leaf transpiration increased with VPD, but transpiration rates slowed above VPDs of 2 kPa.

Both carbon exchange rates and stomatal conductance are reduced in response to high VPD. While the direct impacts of temperature on these effects have already been discussed, changes in relative humidity also alter transpiration, energy balance and tissue temperatures (Conaty *et al.*, 2014). These in turn may affect ion uptake, carbon assimilation, water transport and other processes, each of which has further physiological consequences that confound direct responses to relative humidity (Grantz, 1990).

There is also evidence that increasing VPD can cause inhibition of photosynthesis unrelated to stomatal closure (Bunce, 1983; Morison and Gifford, 1983; Pettigrew *et al.*, 1990). However, Duursma *et al.* (2013) found that in cotton, leaf photosynthesis was relatively insensitive to VPD as it decreased on average only 13% from the maximum leaf photosynthesis over the range of VPD in well-watered conditions. Rawson and Begg (1977) compared the VPD responses of a number of C_3 species, including wheat, soybean, sunflower and sorghum, to step changes in VPD over the range 0.8 to 2.7 kPa. They found little or no response to VPD in these species, which were also under high light and were well-watered. Consequently, high VPD effects on elevating transpiration and small effects on photosynthesis led to lowered instantaneous leaf-level transpiration efficiencies (μmol CO_2 mmol^{-1} H_2O) in cotton (Duursma *et al.*, 2013).

3.1.5. *Effect of drought*

Global warming and climate change will be associated with changes in patterns of precipitation and availability of water; hence, crop plants in some regions may be subjected more often to conditions of drought and plant water deficits. Water deficit is the major abiotic factor limiting plant growth and crop productivity around the world. In all agricultural regions, yields of rainfed crops are periodically reduced by drought, and the severity of the problem may increase due to changing world climatic trends (Le Houérou, 1996). Plant water deficits depend both on the supply of water to the soil and the evaporative demand of the atmosphere.

A recent review by Loka *et al.* (2011) has summarized the effects of water-deficit on cotton growth and physiology, noting that water constitutes a primary component of plant mass and is essential for plant nutrient transport, chemical and enzymatic reactions, cell expansion and transpiration. Water deficits cause morphological and anatomical changes, along with changes in physiological and biochemical responses that affect plant function.

Photosynthesis and respiration

Water is required for photosynthesis at the biochemical level. Light energy is used to oxidize H_2O to produce O_2, NADPH and ATP (Knox *et al.*, 2005). When water stress occurs, it affects the efficiency with which absorbed radiation is utilized in carbon fixation at the leaf level. Plants respond to mild and moderate water stress by closing their stomata, resulting in decreased internal $[CO_2]$ of the leaf (C_i) (Ennahli and Earl, 2005; Massacci *et al.*, 2008). Stomatal conductance is initially more sensitive to the onset of plant water stress than photosynthesis; however, photosynthesis is quickly reduced as plant water deficit becomes more severe (Baker *et al.*, 2007). Similar results have been found in cotton where mild to moderate stress reductions in photosynthesis have been associated with reductions in leaf conductance (Baker, 1965; Baker *et al.*, 2007), leading to a decline in plant growth (Arriaga *et al.*, 2009; Ko and Piccinni, 2009).

The onset of water stress in cotton also increases photosynthetic electron transport. Additional energy is used to increase the rate of photorespiration, while photosynthesis is kept constant or slightly decreases. This process occurs in order to prevent an over-reduction of the photosynthetic apparatus (Massacci *et al.*, 2008). However, under severe deficit water stress, photosynthesis may also be restricted by non-stomatal effects, such as inhibition or down-regulation of metabolic processes at the level of the chloroplast, leading to decreased RuBP regeneration and inhibition of photosynthesis (Flexas and Medrano, 2002; Ennahli and Earl, 2005; Baker *et al.*, 2007). Chloroplast-level effects are typically observed only under very severe stress, where net photosynthetic assimilation is reduced by more than 80% (Ennahli and Earl, 2005). In these studies, re-watering of severely stressed plants can completely reverse the diffusive limitation caused by stomatal closure, but reductions in photosynthesis can continue because of continuation of chloroplast-level inhibition. Loka *et al.* (2011) reported that non-stomatal responses for cotton were varied.

Stomatal conductance and transpiration

Stomatal conductance and transpiration are affected by a complex interaction of factors internal and external to the plant leaf, including soil water availability (Ko and Piccinni, 2009). Continued transpiration results in depletion of soil moisture, and without replenishment can lead to decreases in transpiration rate. Progressive soil moisture deficits trigger root-to-leaf abscisic acid (ABA) signalling through the transpiration stream and induces stomatal closure (Medrano *et al.*, 2002). As available water drops below 60%, evapotranspiration is reduced as a result of slower hydraulic conductance of water to roots, reduced transpiration from stomatal closure, and parahelionastic leaf movements (Hearn and Constable, 1984).

Growth, yield and fibre quality

In many crop species, drought stress-generated reductions in final yield depend on the duration and severity of drought stress, as well as the growth stage at which the stress occurs. Typically in most plants, reproductive growth is more sensitive to plant water deficit than vegetative growth (Baker, 1965), and cotton is similar as the highest water demand occurs during reproductive growth (Hearn, 1980). In cotton cell expansion, cell-wall synthesis and protein synthesis in fast growing tissues (leaf and stem) are among the processes most sensitive to water deficits (Sadras and Milroy, 1996). At the leaf level, stomatal conductance and photosynthesis were less responsive to water deficits than tissue expansion (Sadras and Milroy, 1996). Physiological processes such as stomatal conductance, photosynthesis and respiration are consequently impaired with further implications for metabolic functions such as carbohydrate and energy production, as well as carbohydrate translocation and utilization, while at the plant and crop level, a reduction in individual leaf size consequently reduces leaf area and light interception (Hearn, 1980). Water deficits also increase canopy temperatures, which can contribute to reductions in leaf growth (Pettigrew, 2004b).

Water deficits reduce yield by limiting the production of nodes and fruiting sites, cause fruit shedding and sometimes reduce final boll size and lint produced per seed. Drought will reduce the total dry weight and yield concurrently, but not affect the fractions of total weight in leaves, stems and bolls at the end of the season; however, it will affect the pattern of boll setting (Hearn and Constable, 1984). Pettigrew (2004a) showed that irrigation altered the distribution of bolls both vertically and horizontally on the plants. Irrigated cotton set more bolls at higher plant nodes and further out on the sympodial branches than rainfed cotton. Cotton plants tend to compensate for lack of moisture by shedding fruit and, thereby, to some extent alleviate stress on the remaining fruiting forms (Ramey, 1986). Saranga *et al.* (1998) also showed that drought conditions caused more motes (cotton ovules that fail to ripen into mature seeds) to be produced, which can also contribute to yield reduction.

Since the fibre is primarily cellulose, any influence on plant photosynthesis and production of carbohydrate will have a similar influence on fibre growth. Cell expansion during growth is strongly driven by turgor, so plant water relations will also affect fibre elongation in the period immediately following anthesis. Water-stressed crops usually have fibre with lower final length (Hearn, 1976b; Constable and Hearn, 1981; Ramey, 1986; Pettigrew, 2004b; Lokhande and Reddy, 2014b).

Fibre micronaire, an estimate of fibre fineness and maturity, can be inconsistently affected by drought effects. Depending upon when the stress occurs and its duration, micronaire can either be decreased (Eaton and Ergle, 1953; Marani and Amirav, 1971; Ramey, 1986) or increased (Bradow and Davidonis, 2000). These micronaire variations are thought to be tied to how the drought stress affects the relationship between the photo-assimilate supply (source) and the boll load (sink) (Pettigrew, 1995; Lokhande and Reddy, 2014b). The balance between boll load and crop canopy size can be significant, with high boll loads having lower micronaire, presumably from internal competition (Brook *et al.*, 1992). More recently, Lokhande and Reddy (2014a) found that fibre length, strength and uniformity declined linearly with decrease in leaf water potential (LWP), whereas fibre micronaire increased with decrease in LWP. Fibre strength was most responsive to changes in LWP followed by micronaire, length and uniformity. Immature fibre content increased and fibre maturity ratio declined with diminishing LWP.

3.1.6. Effect of rainfall intensity (flooding/waterlogging)

Waterlogging involves a complex interaction of factors affecting both the soil environment and plant growth. Poor soil drainage, intensive irrigation and highly variable weather patterns are responsible for soil waterlogging affecting crop production. Partial O_2 deficiency in soils (hypoxia) or complete absence of oxygen (anoxia) will damage plant roots and hamper overall plant growth by limiting aerobic respiration (Jackson and Drew, 1984) and nutrient uptake (Setter *et al.*, 2009). There is also a build-up of toxic gases such as CO_2 and ethylene that are generated by the roots and microorganisms which can impair root and whole plant function. The rate of O_2 depletion from waterlogged soils depends on soil temperature and respiration rates of roots and soil microorganisms.

Development of aerenchyma is one of the most common responses in many plant species at the anatomical level, which facilitates oxygen diffusion into root tissues (Jackson *et al.*, 2008). Other morphological changes include increased root porosity via development of adventitious roots and hypertrophied lenticels, and rapid shoot elongation in some waterlogging tolerant species. Modification in water relations, stomatal changes, decreased transpiration and photosynthesis are the physiological responses in plants. Metabolic adaptations including energy production via fermentation, metabolic adjustments and anaerobic protein synthesis are also crucial for survival of plants exposed to low soil O_2 concentration.

Cotton has a root system that is poorly adapted to waterlogged conditions and is rapidly damaged (Leonard and Pinckard, 1946; Huck, 1970; Ellis *et al.*, 2000). Solution experiments show root growth is markedly impaired at O_2 concentrations below about 8% (Leonard and Pinckard, 1946). Complete anoxia causes root growth to cease within 2–3 min and the root tip meristems of all taproots to die within 3 h (Huck, 1970). Cotton does not produce functional aerenchyma under waterlogged conditions (Leonard and Pinckard, 1946; Huck, 1970), and aerobic respiration cannot proceed. The plants rely on anaerobic fermentation, leading to the production of ethanol, which enables the plant to maintain ATP production albeit with a reduced energy yield. Two key enzymes in the alcoholic fermentation pathway are pyruvate decarboxylase (Pdc) and alcohol dehydrogenase (Adh) (Drew, 1997). However, in cotton endogenous levels of Adh are low (Ellis *et al.*, 2000).

As a consequence of these effects, there have been reported reductions in leaf area and photosynthesis with waterlogging leading to restrictions in overall cotton growth and fruiting development resulting in reduced lint yields (Hodgson and Chan, 1982; Bange *et al.*, 2004). Waterlogging results in decreased photosynthesis in cotton (Meyer *et al.*, 1987; Guo *et al.*, 2010), and variation amongst cotton genotypes has been reported (Conaty *et al.*, 2008). Milroy and Bange (2013) were able to demonstrate that there was some association with photosynthesis of the most recent fully expanded leaf and its nitrogen (N) content, but it could not fully explain reasons for reduction in photosynthesis from waterlogging. Nitrogen concentrations in these

17

leaves are reduced with waterlogging (Milroy *et al.*, 2009). This suggests that other mechanisms, in addition to those acting through the reduced uptake of N, were likely to be acting on leaf performance. Recent studies by Najeeb *et al.* (2015a) have demonstrated that along with reductions in leaf N content there were also reductions in photosynthesis associated with increases in leaf ethylene concentrations. Reductions in stomatal conductance were also reported by Najeeb *et al.* (2015a). Consistent with these findings, reductions in crop-level measurements of crop radiation use-efficiency, canopy leaf growth and total final biomass production have also been observed (Meyer *et al.*, 1987; Sahay, 1989; Bange *et al.*, 2004).

Reported yield losses in cotton under waterlogging have been mainly associated with reduced boll number and small changes in boll size (Hodgson, 1982; Hodgson and Chan, 1982; Bange *et al.*, 2004). All of these studies consistently found reductions in boll number in response to waterlogging, but boll size was less responsive and in some cases did not change significantly. Bange *et al.* (2004) reported that this reduction in boll number was commensurate with the reductions in total plant dry matter due to lower radiation use efficiency. Results from their studies also suggested that this reduction in boll number is most likely associated with reduction in fruiting site production rather than increased shedding alone. In the same study, these effects were only measured earlier in the flowering period, a finding supported by Reicosky *et al.* (1985). There is also some evidence that there is a degree of acclimation associated with repeated waterlogging events (Reicosky *et al.*, 1985; Milroy and Bange, 2013).

As suggested, waterlogging cotton has also been associated with fruit shedding resulting in lower fruit retention and boll numbers measured at harvest (Conaty *et al.*, 2008; Najeeb *et al.*, 2015b). Ethylene induction in cotton plants is associated with a wide range of injuries and stresses (Christianson *et al.*, 2010) and can result in increased rates of abscission of young fruit (squares and bolls) (Guinn, 1974). However, its involvement with the impact of waterlogging on cotton has not been widely researched. Aminoethoxyvinylglycine (AVG), a known ethylene inhibitor (Jackson, 1985), has been used in cotton and increased cotton yields under waterlogging (Bange *et al.*, 2010c; Brito *et al.*, 2013) and been associated with increased fruit retention at harvest (Najeeb *et al.*, 2015b). Further, more detailed studies have shown that ethylene concentrations increased (especially in waterlogging sensitive genotypes) in leaves and young fruit when exposed to severe waterlogging (Najeeb *et al.*, 2015a). At the molecular level, Christianson *et al.* (2010) also found changes in the level of expression of genes associated with ethylene in cotton after waterlogging. These findings emphasize the potential role of ethylene in mediating responses to waterlogging.

In addition to the physiological impacts of waterlogging on the crop, there are also significant impacts on nutrient availability and uptake. The availability of N, Fe and Zn (reduced) and Mn (increased) in the soil are directly affected by the decline in soil oxygen, and uptake of N, K and Fe by the roots is also impaired (Milroy *et al.*, 2009). There have been very few reported impacts of waterlogging on fibre quality. Bange *et al.* (2004) found no effect, while Najeeb *et al.* (2015b) found small reductions of both fibre length and micronaire.

3.1.7. *Interactive effects of climate change*

Combined temperature and carbon dioxide effect

Global climate models project that rising atmospheric [CO_2] may cause an increase in both global average temperatures and in temperature extremes (IPCC, 2014). Interactions between [eCO_2] and temperature are complex, thus it is necessary to consider the impact that both warmer temperature and [eCO_2] combined will have on cotton physiology and growth.

Reddy, A.R. *et al.* (1998) showed that photosynthetic rates of cotton increased with [eCO_2] and warmer air temperatures. However, the response of stomata to increased [CO_2] and warmer temperatures is variable. An experiment conducted on cotton grown in controlled environment chambers showed that [eCO_2] reduced stomatal conductance at all three temperature (26/18°C, 31/23°C and 36/28°C) treatments, but the magnitude of the reduction depended on growth temperature, with the greatest reduction of stomatal conductance with [eCO_2] occurring at 31/23°C (Reddy, A.R. *et al.*, 1998). In experiments conducted by Reddy, V.R. *et al.* (1995), canopy transpiration rates of plants in [eCO_2] (700 µmol mol^{-1}) were lower than in ambient [CO_2] (350 µmol mol^{-1}) in all temperatures grown (20/12°C, 25/17°C, 30/22°C, 35/27°C). However, canopy transpiration increased with increasing temperature, despite the reduction in transpiration at [eCO_2] (Reddy, V.R. *et al.*, 1995) indicating that [eCO_2] may not negate the negative impact of warmer temperatures on plant water use of cotton. Although the higher photosynthetic rates and the slightly lower transpiration rates of plants grown in high [CO_2] resulted in greater leaf-level water use efficiency, the actual quantity of water required by cotton plants in these environments may be increased, but this is yet to be determined. Hence, we cannot assume that the responsiveness of photosynthesis and growth to [eCO_2] will become greater with global warming (Stokes and Howden, 2010).

There have been a number of SPAR studies to assess the combined effects of [eCO_2] and warmer temperatures on fruit production and retention of cotton. Reddy *et al.* (1999) reported that cotton grown in high atmospheric [CO_2] (720 µmol mol^{-1}) produced more squares and bolls, because additional vegetative growth was associated with greater photosynthesis. This is based on higher photosynthetic rates and greater leaf area leading to the higher production of assimilates used in metabolic sinks, such as reproductive structures. Over a wide range of temperatures, [eCO_2] increased the number of fruiting organs and the retention of bolls (Reddy *et al.*, 1999); however, Reddy, K.R. *et al.* (1998) found that although more fruiting sites were produced at 700 µmol mol^{-1} [CO_2] for all temperatures, fruit retention at 32°C and 36°C was lower than at 27°C, thereby suggesting that it is unlikely that [eCO_2] will ameliorate the effect of high temperatures on flower abortion. Therefore, it is still unclear as to what the net outcome of warmer temperatures and elevated [CO_2] will be for fruit retention, and ultimately yield, for cotton grown in projected future environments.

Combined CO_2 and water stress effects

Changes in atmospheric [CO_2] may alter water use and water use efficiencies in plants, but conversely the effect of [eCO_2] on plant physiology and growth is likely to be influenced by plant water availability. In a SPAR experiment, Ephrath *et al.* (2011) showed that cotton grown in an [eCO_2] treatment (700 µmol mol^{-1}) used less water than cotton grown in ambient CO_2 treatments (350 µmol mol^{-1}) in both well-watered and water-stressed conditions. This is attributed to lower stomatal conductance of cotton grown at [eCO_2] compared with plants grown at ambient [CO_2], resulting in lower transpiration rates (Reddy, A.R *et al.*, 1998; Ephrath *et al.*, 2011). In contrast, glasshouse experiments have demonstrated that cotton grown at [eCO_2] (710 µmol mol^{-1}) had higher plant water use than ambient (352 µmol mol^{-1}), which was attributed to increased leaf area, and thus more rapid depletion of soil moisture as the canopy developed (Samarakoon and Gifford, 1995, 1996). However, FACE experiments in Arizona during 1990 and 1991 showed that there were no differences in evapotranspiration of cotton grown at [eCO_2] (550 µmol mol^{-1}) compared with ambient [CO_2], determined using three different methods (Dugas *et al.*, 1994; Hunsaker *et al.*, 1994; Kimball *et al.*, 1994). These reported differences may also be due to pot and soil bin effects, variety and other environmental effects and thus these concepts should be tested further in both glasshouse and field environments to improve understanding of water use of cotton in high [CO_2] environments.

3.2. Climate Change Impacts on Pests and Diseases

Insects, weeds and diseases are recognized threats to cotton production worldwide. Their growth, phenology and geographical distribution (including effects on hosts) with future climate change will be affected by warmer temperatures, changes in rainfall distribution and intensity, changes in wind patterns and frequency of extreme weather events.

The impact of temperature has been identified as a dominant factor that affects herbivorous insects with little direct effects of [CO_2] (Sankaranarayanan *et al.*, 2010). It has been suggested that most insects can adapt their body temperature to that of the environment, and that global warming is slow enough for insects to adapt. As a consequence, it is expected that insect pests affecting cotton production will live and thrive in future conditions (ITC, 2011).

Temperature directly affects insect development, survival, number of generations, timing and the duration of diapause, while rainfall affects the growth of plant hosts leading to differences in distribution and abundance of insect pests. Gregg and Wilson (2008) suggested the impacts of rising temperatures were increased abundance of some insects in new and existing regions (they used silver leaf whitefly (*Bemisia tabaci* B Biotype) as an example), increased development rates of some insects and increased overwintering of some pests. They cautioned, however, that these effects of warmer temperatures on insects could be offset by increased insect mortality. Furthermore, overwintering is influenced more by the amount of rainfall in autumn and winter (which affects the growth of plant hosts) than by temperature *per se*.

The relative effect of warmer temperatures on the growth rates of crops and insects together also needs to be considered. In non-stressed conditions crop growth rates may increase in line with increased rates of insect development, potentially negating the effects of increased insect development and damage to the crop. An example of this was given by Heagle (2003), where the effects of thrips on leaf scarring was increased under [eCO$_2$]; however, the effects were negated with increased leaf area which was associated with [eCO$_2$].

Research into [eCO$_2$] has also shown that it can affect leaf chemistry, which may have effects on insect feeding patterns and survival (Sankaranarayanan *et al.*, 2010). Host plants grown in [eCO$_2$] exhibited increased biomass with increased C/N ratio (+27%), decreased N content (−16%) and increased concentrations of tannins (+29%) and other phenolics (Heagle, 2003). While it has been suggested that these changes in leaf chemistry may increase insect mortality because insects need to consume more plant material, Coviella *et al.* (2002) found that these effects only occurred with low plant N nutrition in [eCO$_2$].

Coviella *et al.* (2000) and Wu *et al.* (2007) have also shown that cotton plants grown in [eCO$_2$] (doubling of ambient) have less expression of the *Bacillus thuringiensis* (Bt) toxin in leaves. This effect did not change the performance of the cultivars to resist *Helicoverpa* spp. as there was evidence to suggest that *Helicoverpa* were adversely affected by feeding on cotton subjected to [eCO$_2$]. While their lifespan increased, their pupal weight, survival rate, fecundity, frass output, relative mean growth rates and the efficiency of conversion of ingested and digested food was decreased (Wu *et al.*, 2006; Chen *et al.*, 2007).

In the case of weeds, there are concerns that some weed species will increase their competitive advantages under [eCO$_2$] (IPCC, 2007a). Potentially there will be direct and indirect consequences affecting fitness of weeds and crop and altered weed–crop competitive interactions (Sankaranarayanan *et al.*, 2010). Ziska *et al.* (1999) have also suggested that it may increase tolerance to glyphosate in some weed species.

Disease control may be affected by warming and [eCO$_2$] by creating environments that are more conducive to disease proliferation and affecting the ability of the host plant to resist disease (ITC, 2011). It is also most likely that regions that have future climates that result in higher humidity will be more susceptible than those with drier climates. For all pest and disease situations, chemical control methods may sometimes become less efficacious due to faster decomposition of chemicals under higher temperatures (ITC, 2011), changes in leaf surface characteristics and accumulation of starch in leaves (Sankaranarayanan *et al.*, 2010).

3.3. Climate Change Impacts on Soils

Cotton production, particularly high-yielding irrigated cotton, relies on a large amount of N fertilizer to maximize yield. Changes in crop performance induced by climate change (e.g. biomass production, quantity and quality of organic input to the soil, water and nitrogen use efficiencies) can lead to changes in plant N requirements, water and N uptake and N cycling in the soil. Climate change itself can have direct impacts on soil processes that determine the availability of N for plants through changes in microbially-mediated processes such as N mineralization, immobilization, nitrification and denitrification (Fig. 1). While the number of studies on climate change impact on soils in cotton system is very limited, knowledge on how climate change impacts soil processes can provide some insights into its potential impacts on cotton soil.

3.3.1. Effect of elevated CO$_2$ concentration

[eCO$_2$] increases C input belowground (e.g. Luo *et al.*, 2006), and this has also been observed in cotton systems (Prior *et al.*, 1994; Wood *et al.*, 1994). Such increases stimulate microbial activity (Runion *et al.*, 1994) thereby enhancing microbially-mediated soil processes. Studies have found that [eCO$_2$] impacts on C mineralization (Korner and Arnone, 1992; Hungate *et al.*, 1997b; Norby *et al.*, 2004; Jackson *et al.*, 2009), N mineralization and nitrification (Hungate *et al.*, 1997a; Langley *et al.*, 2009; Muller *et al.*, 2009) and denitrification (Carnol *et al.*, 2002; Baggs *et al.*, 2003), although the magnitude and direction of its effect varies among the studies. In general, it has been found that belowground responses to [eCO$_2$] are often greater than aboveground responses (Jackson *et al.*, 2009), however, in some cases, rates of litter decomposition may slow down due to the preferential use of labile-C over complex-C by soil microbial communities, which in turn may lower CO$_2$ emissions and favour C sequestration in the soil.

It is also argued that such changes in belowground C supply may lead to immobilization of soil N in the long term, thereby limiting the N available for plants and creating a negative feedback that constrains future increases in plant growth (Diaz *et al.*, 1993; Luo *et al.*, 2004). A review by Hu *et al.* (2006) concluded that the microbial control of N availability under [eCO$_2$] depends on the initial ecosystem N status and the nature and magnitude of external N inputs. Given the

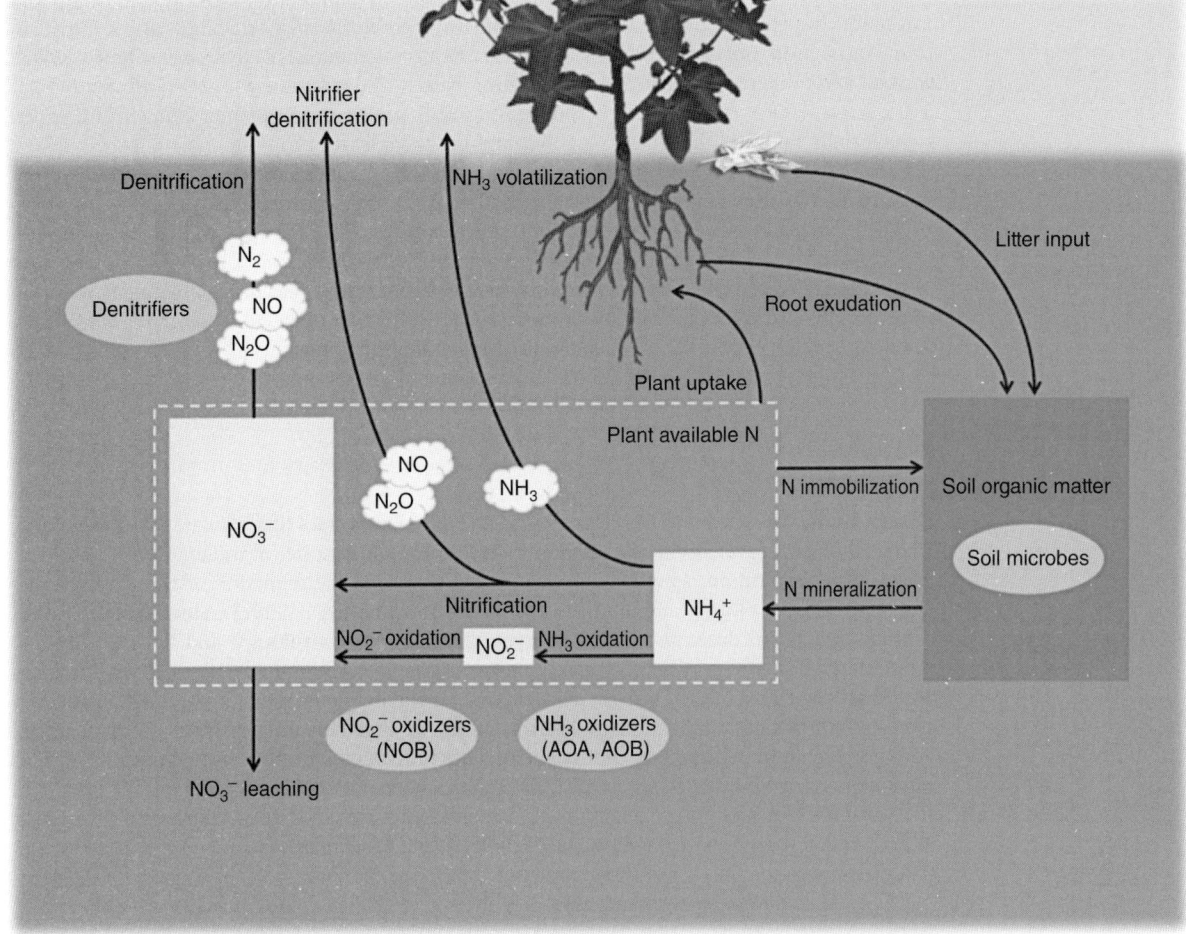

Fig. 1. Soil N-related processes that are likely to be influenced directly or indirectly through changes in crop response to climate change. AOA, ammonia-oxidizing archaea; AOB, ammonia-oxidizing bacteria; NOB, nitrite-oxidizing bacteria.

highly managed nature of cotton production systems (both water and fertilizer management), understanding the long-term effect of [eCO$_2$] on soil processes is critical in identifying the best management regimes and developing the adaptation strategies to mitigate the negative impact of [eCO$_2$] on soils that may impact cotton productivity.

There is a plethora of evidence for the impact of [eCO$_2$] on the composition of soil microbial community in natural systems (e.g. Blankinship *et al.*, 2010). While the number of studies on agricultural systems is relatively limited, several studies have shown that [eCO$_2$] significantly modified soil microbial biomass (Anderson *et al.*, 2011) and functional composition and structure of soil microbial communities (i.e. ammonia oxidizers and denitrifiers) in agricultural systems (He *et al.*, 2014). Changes in soil microbial community can alter soil processes that regulate C and N cycling in the system, thus studies on the impact of [eCO$_2$] on soil processes need to incorporate both soil processes and microbial responses to [eCO$_2$] and the feedback between soil and plant responses to [eCO$_2$].

3.3.2. Effect of elevated temperature

Temperature influences many biological, physiological, chemical and physical processes that occur in soil. Elevated temperature has been shown to increase soil respiration (Pendall *et al.*, 2004; Davidson and Janssens, 2006), and an increase of 2°C in the global average temperature is predicted to increase soil C release by 10 Pg, mainly owing to increases in microbial activity (Singh *et al.*, 2010). Microbial growth and activity are often regarded as temperature-sensitive, as rates of enzymatic reactions are primarily dependent on temperature. Not only changes in the activity of the soil microbial community but also shifts in the composition of the microbial community have been observed under soil warming, with the fungal community often more stimulated than the bacterial community (Zogg *et al.*, 1997; Zhang *et al.*, 2005).

Soil N-related processes are also influenced by elevated temperature. Studies have found that in general, warming significantly increased net N mineralization, nitrification and denitrification (Aerts, 2006; Bai et al., 2013), with its effect more pronounced in a cold environment which showed a 70% increase in net N mineralization with a mean warming of 2.4°C (Rustad et al., 2001; Aerts, 2006). However, the responses of soil N processes to elevated temperature varied substantially between the studies, highlighting the complexity of interactions between abiotic (e.g. soil moisture) and biotic (e.g. microbial community, plant uptake) controlling factors that are modified by elevated temperature (Barnard et al., 2005; Bai et al., 2013).

While temperature can directly impact soil processes, indirect effects of temperature through plant responses can significantly impact soil processes that control C and N cycling. An accelerated rate of crop development at elevated temperature (see Section 3.1.3) means that the timing/duration of crop cultivation, rotation and resource management may be changed. Such changes are likely to have a greater impact on soil C and N cycling within the system.

3.3.3. Effect of drought

Drought can have direct and indirect impacts on key soil processes that regulate ecosystem nutrient cycling. Water availability is a dominant determinant of the rates of mineralization, nitrification and denitrification, as water deficiency often results in the reduction of these microbially mediated processes (e.g. Stark and Firestone, 1995; Hartmann et al., 2013). Simulated prolonged drought conditions decrease potential enzymatic activities involved in C and N cycles, thereby reducing the rate of C and N mineralization via its direct effect on soil water content (e.g. Sardans and Penuelas, 2005, 2010). Drought can also impact soil processes indirectly through changes in plant growth and activity that regulates the supply of organic substrate for microbial activity. This was highlighted in a study conducted in a large cotton-producing area in Texas, USA. Following the record drought/heatwave in 2011, the study found that key enzymes involved in C and P cycling were in fact higher during peak drought/heatwave conditions than 8 and 12 months after the peak period (Acosta-Martinez et al., 2014b), suggesting that recovery of soil water content alone may not be sufficient to recover soil enzymatic activities.

Drought also impacts the composition of the soil microbial community (Berard et al., 2011; Acosta-Martinez et al., 2014a), as microbial communities respond to moisture levels directly, as they require water for physiological activities, and indirectly, owing to the effect of changing soil moisture on gas diffusion rates and oxygen availability (Singh et al., 2010). Drought often favours fungal populations relative to bacterial populations (Bell et al., 2009; Bapiri et al., 2010; Acosta-Martinez et al., 2014a), owing to their extensive hyphal network to access water. Such differences in drought responses among different microbial groups are frequently reported (Fuchslueger et al., 2014; Keil et al., 2015).

The recovery of soil from the drought conditions, i.e. rewetting of soil following drought, can induce a pulse of soil respiration, N mineralization and denitrification, with the magnitude of this effect increased with increasing intensity of soil drying and rewetting (Guo et al., 2014). However, studies have also found that the recovery of microbial growth and their functions may be slower (Goransson et al., 2013), particularly with a prolonged drought period (Meisner et al., 2015). This could have an important implication for irrigated cotton where the current practice includes a dry-down phase between the seasons to improve soil structure before filling the soil profile with water prior to the planting.

3.3.4. Effect of rainfall intensity (flooding/waterlogging)

Flooding can cause a severe loss of soil N through N leaching and denitrification. In well-aerated soils, nitrification is a more dominant process than ammonification; however, in anaerobic conditions nitrification becomes inhibited (Unger et al., 2009). Furthermore, denitrification becomes more dominant under anaerobic conditions, thus flooding can cause a shift in the relative amount of ammonium and nitrate. Nitrate is highly mobile in soil, thus it can also be leached from the system under flooded conditions. Previous studies have found that flooding caused soil nitrate concentration to decrease and soil ammonium concentration to increase, as expected under anaerobic soil conditions (Unger et al., 2009).

Oxygen is used as an electron acceptor in an aerated soil; however, other compounds are used in an anaerobic, waterlogged soil. Denitrification occurs when nitrate is used as an electron acceptor by denitrifying micro-organisms, resulting in a rapid loss of nitrate in the soil. Waterlogging not only increases N losses from the soil, it also increases N_2O emission through denitrification.

N_2O produced as a by-product of denitrification processes is a GHG 298 times more potent than CO_2, and it is estimated that agriculture contributes approximately 58% of total anthropogenic N_2O emission (IPCC, 2007b). Irrigated cropping systems with a large N fertilizer input are particularly susceptible to N_2O emissions, particularly during heavy rainfall events (Harris *et al.*, 2013). Several studies have estimated that only ~1% of applied N is lost as N_2O during the cotton-growing season from flood-irrigated alkaline clay soils in Australia (Rochester, 2003; Grace *et al.*, 2010). This is largely due to the high soil pH of the studied system, as alkaline soils produce relatively less N_2O than N_2 during denitrification. In fact, it is estimated that approximately 16% of applied N is lost through denitrification as N_2O and N_2. Furthermore, recent studies have identified the occurrence of bacteria responsible for anaerobic ammonium oxidation (anammox) – a process that converts ammonium and nitrite into dinitrogen in agricultural soils (Humbert *et al.*, 2010; Long *et al.*, 2013). Thus, the projected increase in the frequency and intensity of floods under climate change is likely to increase the occurrence of waterlogged/anaerobic soil conditions, further exacerbating the loss of N through the system.

Intense rainfall can also cause a loss of N through leaching of nitrate below the rooting zone of cotton. Irrigated cotton systems suffer from a substantial loss of N, with a leaching of over 200 kg ha^{-1} year^{-1} of nitrate being reported from an irrigated Vertosol in Australia (Moss *et al.*, 1999; Hulugalle *et al.*, 2005a). Unlike ammonium, which is relatively immobile in soil (as it is positively charged and hence attached to negatively charged clay particles and organic matter), nitrate is water-soluble and susceptible to downward movement in soil upon heavy rainfall.

Cotton is intolerant of waterlogging, exhibiting both physiological and nutritional symptoms upon waterlogging events (Bange *et al.*, 2004; Milroy and Bange, 2013, see Section 3.1.6). Significant impacts of rainfall intensity on soil N availability suggest that crop response to waterlogging can also be attributed to changes in soil nutrient availability, as waterlogging increases N losses through denitrification and nitrate leaching.

IV MANAGEMENT APPROACHES TO ADAPT TO IMPACTS OF CLIMATE CHANGE

National yields of cotton vary widely, but in some countries yields have been increasing in the past three decades. In general, those countries with higher yield have been increasing at rates from 10 to 30 kg lint ha^{-1} year^{-1} (e.g. USA, Australia, Brazil and Turkey), but there are a number of countries and regions such as Uzbekistan and West Africa where production systems have stagnated in yield, probably due to combinations of reduced water availability and soil constraints, including crop nutrition.

Contributing to the variability in performance is the 'yield gap' that exists between the average yields attained by an industry or region compared to yields recorded in regional crop competitions (yield potential). The factors responsible for yield gaps vary widely for a particular field or system and identifying the limiting factors is an important challenge in order to correct those factors and reduce the yield gap. In a recent review by Constable and Bange (2015), they suggest that the greatest reduction in the yield gap between farm averages and yield potential will be more likely through removing yield constraints of the poorest fields and systems. This may include crop rotation and tillage systems which improve fallow moisture or improved nutrition. If lepidopterous insects or weeds are a major constraint for yield, then transgenic insecticide (e.g. Bt) traits or herbicide resistance traits have helped to improve yield (Campbell *et al.*, 2014). Likewise, where disease is a constraint, resistant cultivars are important (Liu *et al.*, 2013). It is possible that challenges that currently restrict realization of yield potential will, in many instances, be amplified with future climate change.

Conversely, in some situations climate change may raise yield potential. In the same review by Constable and Bange (2015), they re-evaluated an estimate of theoretical yield of cotton to be about 5000 kg lint ha^{-1} and showed that yields in irrigated systems were attained in the range of 3500 kg ha^{-1}. The increase in yield over the Baker and Hesketh (1969) estimate (4355 kg lint ha^{-1}) is probably due to [eCO$_2$], longer growing season and improved cultivars (particularly with increased lint fraction and harvest index). A key point noted in this review was that to attain higher yields, a longer season was required than currently used in commercial practice. It is possible that in a future climate, where season length may be increased, management of major biotic and abiotic stresses by growers could increase yield potential and approach estimates of theoretical yield.

Although the rate of climate change is relatively slow, a small change in average temperature may have a relatively large impact. For example, with days of 25°C maximum and 15°C minimum temperature, it takes about 63 days from planting to produce first-square; adding 1°C to the mean temperature will decrease the number of days to first-square by 7 days. Even so, the rate of climate change is still relatively small compared with the year-to-year variability in temperatures experienced in many cotton cropping regions.

Therefore, practices that are adopted on the farm to deal with climate and environmental variability and improve sustainability offer significant opportunity to assist with climate change. The wide geographic spread of production also means that some management practices differ as growers have adapted their practices for their various climates. It will be important in developing adaptation strategies to consider a range of options that have synergistic effects in reducing the impact of climate change (e.g. cultivar, frequent irrigation, planting date strategies), and avoid relying on a single approach. In addition, management options will need to be flexible to cope with potentially more variability and uncertainty, and planned to avoid creating problems (mal-adaptation) elsewhere in the farming system (e.g. pest resistance to pesticides).

Carberry *et al.* (2011) highlighted the need to work with growers to identify practices to improve efficiencies with little added risk, as well as improving the 'efficiency frontier' by adopting new production systems and/or practices (utilizing new technologies such as biotechnology and precision agriculture) which can increase returns with little added risk or investment. A considerable role for research and extension exists and many initiatives are in place to achieve these strategies. The front line of defence for producers will be an evaluation and appropriate alteration of their production practices. In this section, we will outline practices that can be used to build resilient cotton systems in the face of climate change. The role of research in assisting these efforts will be specifically addressed in a later section of this review.

4.1. Cultivar Change

Cultivar choice is a strong component of realizing both target yield and fibre quality levels on farm. A balance needs to be reached between yield, fibre quality/price and other important considerations such as disease resistance, and insect and herbicide resistance. Most cotton breeding programmes throughout the world are developing cotton cultivars well suited to the environmental and climatic conditions experienced in that particular cotton production region (hot and cool), and therefore give growers options for selecting cultivars suitable to their environment. There are also specific breeding programmes in place for rainfed cotton production (Stiller *et al.*, 2005). This means that there are already cultivars available to cotton producers that can provide some resilience to climate change.

For cotton producers, there are a number of key considerations described below that will be useful as the climate changes over the coming decades. In regions where there are significant risks of abiotic (e.g. high temperature, waterlogging) and biotic (e.g. insect, diseases) stresses, cultivars that demonstrate more resilience to these stresses should be considered. Brown and Oosterhuis (2010) showed that obsolete cultivars were more resilient to heat stress than modern commercial cultivars. In the future, it will be important for cultivars to be differentiated for their resilience to heatwaves (Bibi *et al.*, 2008b) and waterlogging (Conaty *et al.*, 2008), in the same way cultivars are promoted for their disease tolerance. Similarly, there will be a need for considering cultivars for their relative efficiencies in resource use (water and nutrition). Universal descriptions of these tolerances and efficiencies will also be a major factor assisting cultivar selection, especially when a region is serviced by a number of breeding companies.

Season length in a region will be a major factor in cultivar choice. Across regions there are potentially a number of factors that define season length: growing season length determined by thermal dependence (growth influenced by temperature), season length dictated by the occurrence of terminal drought (e.g. Texas High Plains, USA), or an active choice to grow cotton over a shorter season to preserve or utilize limited resources (e.g. water) or reduce input costs (e.g. fertilizers, pesticides). In regions that have production associated with long seasons with adequate resources, there is a preference for more indeterminate and late maturing cultivars. In regions with a shorter season or when terminal drought may occur, there should be a preference for cultivars that are likely to be more determinate, or have growth characteristics that lead to earlier maturity (e.g. shorter phenological stages, higher fruit growth rates) (Bange and Milroy, 2004), or are more responsive to management practices that promote earlier maturity. Cultivars vary in their responses to limited water (Stiller *et al.*, 2004) and the use of growth regulators such as mepiquat chloride (anti-gibberellin) may be considered, but they may lead to differences in crop maturity (Constable, 1995). In rainfed situations with high climate variability, long season growth, and do not necessarily suffer from terminal drought (e.g. Australia), it has been found that cultivars that are less determinate are preferential in delivering higher yields (Stiller *et al.*, 2005) because of their ability to regrow after significant stress events.

There are also choices between cultivars in terms of fibre quality. Growing on a smaller land area, but planting a higher value product, may provide a greater return during periods of limited water. Choosing a cultivar with specific fibre properties can mitigate some climate or management challenges. Examples include longer fibre cultivars reducing short fibre discounts in water-stressed environments, or high micronaire cultivars minimizing low micronaire discounts as a result of cool or stress environments during boll maturation.

At a whole-farm level, a key strategy is to select cultivars that have different adaptive traits to spread risk to variable climate and accommodate changes in management. Consideration may be given to cultivars that minimize impacts of water stress or crop maturity when season length is ill-defined. In their review of climate change focused on Indian production, Sankaranarayanan *et al.* (2010) presented one adaptation strategy where cotton species are mixed on the farm. An intercropping study of a mix of 25% *G. arboreum*, 25% *G. herbaceum* and 50% *G. hirsutum* species at Kovilpatti (Tamil Nadu) in India was found to produce the highest yield and stability under both high and low rainfall situations. More research is needed to establish the yield potential levels at which these strategies provide economic benefit.

4.2. Season Length and Planting Date

One of the consequences of warmer temperatures associated with climate change is an extension of the growing season length. Increases in season length due to temperature occur when seedbeds are warmer earlier, accompanied with less extreme cold events at the start of the season, and where at the end of the season cold temperatures that affect harvest preparation are delayed. These increases in season length have the potential to increase yield. A recent study of predicted climate change effects in 2030 across all Australian cotton production regions predicted that growing season length may increase. This was assessed by showing that the risk of crop stress associated with cold weather (days <11°C) early in growth was diminished, and that the date of last effective flower contributing to lint yield was delayed due to the later average time of first frost/freeze (Luo *et al.*, 2014). A number of studies have shown that yield can increase with a longer season at 14 to 19 kg lint ha^{-1} day^{-1} (Bange and Milroy, 2004), 20 kg lint ha^{-1} day^{-1} (Constable *et al.*, 1976) and up to 34 kg lint ha^{-1} day^{-1} (Stiller *et al.*, 2004).

However, there are some additional challenges required to exploit a longer growing season for improved yield in future climates. With hotter temperatures early in the season, shorter periods of vegetative growth might not be able to support high fruit loads because reproductive development (squaring and flowering) will be quicker. Proper selection of cultivars and management will be required to avoid 'cutout', which will reduce yield potential and negate the potential benefit associated with increased season length; the role of [eCO$_2$] to compensate for these effects is still unknown (Reddy, K.R. *et al.*, 1996). Yeates *et al.* (2010) investigated the effects of early-season irrigation on vegetative growth and found that more vegetative growth was necessary to support high yielding cotton systems that used transgenic cotton with high and early fruit loads. In a recent analysis, Constable and Bange (2015) reaffirmed the need to have continued vegetative growth during early boll set, to allow crops to 'cutout' and mature later, and achieve higher yields. They suggested using a management strategy that regulated vegetative and reproductive growth through the use of water, fertilizer and growth regulators. Overall, increased growing season length may grow more cotton lint, but it will require more resources.

Other factors contributing to the challenges associated with allowing crops to exploit season length may include a loss of reproductive capacity due to reduced boll filling periods, and an increase in fruit shedding arising from extreme high temperature events. Luo *et al.* (2014) found that in future Australian climates there would be shorter boll periods, and this would potentially lead to smaller boll sizes (Reddy *et al.*, 1999). Reddy *et al.* (1997a) also suggested that one of the main losses in reproductive capacity in cotton with elevated temperatures associated with climate change was increased fruit abscission. Wilson *et al.* (2003) showed that with a longer season length and when crop growth is maintained, loss of fruit can be compensated to some degree with new growth, although increased season length coupled with warmer temperatures will require greater resources (water and nutrition) to achieve similar or higher yield outcomes (Constable and Bange, 2015).

One of the clear management benefits associated with longer season length is the potential to provide cotton producers with flexibility regarding planting time decisions. Changes in planting time can offer a 'systems solution' that can provide benefits in maintaining or improving yield, fibre quality and resource use efficiencies. Recent planting time studies into the effects of high and early fruit loads associated with production of Bt cotton provide insights on the opportunities associated with flexible planting time in environments with longer season lengths that may exist in a future climate (Bange *et al.*, 2008; Braunack *et al.*, 2012). Cotton containing genes expressing insecticidal proteins offers control of major lepidopteran pests (particularly *Helicoverpa* spp.), and has allowed for more informed knowledge on the impacts of management and environment on cotton growth without the confounding effects of crop insect damage.

Research by Bange *et al.* (2008) in Australia showed that crops with higher fruit retention can maintain yield and improve fibre length and micronaire for delayed planting dates in warmer and longer seasons. In these studies, yield was maintained for plantings up to 20 days later than the normal planting date as early growth was more rapid when crops were planted into warmer temperatures. In addition, improvements in fibre length were measured and there were reductions in micronaire, both contributing to improved fibre quality. This was associated with the cooler conditions during the early boll-filling stages of the crops. It was noted that planting crops into warmer conditions also had the benefit of avoiding low temperatures at emergence, which can reduce cotton seedling vigour and lead to poor establishment, poor early growth and increased risk of seedling diseases (Oosterhuis and Jernstedt, 1999; Dong *et al.*, 2006; Bange *et al.*, 2008). However, in more determinant seasons such as in the Mississippi river delta (USA), late planting can decrease yields, and in southern Texas late planting can push the crop into the drier part of the season and the start of the hurricane season with adverse effects on yields.

While there was no evidence to support the idea that later planting benefited yield through increased plant size at flowering in the studies by Bange *et al.* (2008), there were improvements in yield with Bt cotton in Mississippi. Pettigrew and Adamczyk (2006) found benefits from early sowing due to an extended growing season leading to higher LAI and greater fruit numbers. Braunack *et al.* (2012) was also able to show that in some cotton-growing regions in Australia, WUE could be improved with later planting. This was associated with maintaining yields while concurrently reducing water use with a shorter growing season and avoiding seasonal growth in periods of high temperatures and low humidity that lead to significant increases in crop water use. Modifying planting time may overcome climatic challenges affecting yield and quality, and harness the benefits of a longer season length.

4.3. Pest Management

One major production issue that growers face each season is the protection of the crop against a range of insect and mite pests. To control these pests, cropping systems have historically relied on intervention with chemical pesticides, which remain a significant component of the cost of production (Fitt and Wilson, 2000). In addition, the use of chemical sprays gives rise to ecological problems arising from pesticide resistance in key pests and environmental concerns about pesticide movement off-farm (Fitt, 2000; Wilson *et al.*, 2004).

The development of genetically engineered (transgenic) cotton with *Bacillus thuringiensis* (Bt) genes have been made available to cotton growers throughout the world. Cotton germplasm containing these genes offers significant potential to reduce pesticide use for the control of major lepidopteran pests (particularly *Helicoverpa* spp.). However, as the system is changing, pests formerly suppressed by these sprays for *Helicoverpa* spp. are emerging as new challenges. As a consequence of improved insect control, retention of squares (flower buds) and young bolls is higher in these crops in some regions, thereby requiring management practices such as planting time to be re-evaluated (Dong *et al.*, 2006; Hofs *et al.*, 2006; Pettigrew and Adamczyk, 2006). Cotton growers can also employ the use of transgenic cotton that allows over-the-top application of herbicides for weed control to enable a rapid response to weed infestations, but this can predispose the system to resistance if not practised with integrated weed management that includes soil residual herbicides, farm hygiene and tillage.

Seasonal climate variability (especially in relation to variations in temperature and rainfall) also influences the distribution and abundance of insects, weeds and diseases. As climate changes, as has occurred in the past, these elements will need to be considered within and between seasons to ensure effective and sustainable management. Pests respond strongly to climate signals and their impacts are highly dependent on climate variability; therefore, lessons for adaptation are manifested in understanding the responses of pests to varied scenarios. Most industries throughout the world already have experienced extremely wet and very dry years, and the pest issues associated with these climates. Wet and warm years can see abundant winter and summer weed hosts contributing to insect pest build up. Hot dry years may have fewer insect pests, which may mean reduced pest control is required throughout the season. Building on the concepts presented in Howden *et al.* (2010), the important pest management strategies for cotton that could be considered to respond to or override climate variability and change are:

- Continued crop improvement to create insect-, disease- and herbicide-tolerant cultivars through conventional plant breeding or genetic modification.
- Implementation of effective integrated insect, weed and disease management that aims to encompass all farm management techniques. Wilson *et al.* (2004) outlined key principles of an insect integrated pest management (IPM) approach, which included practices in the growing season and the 'off season'. The overall goal was to produce profitable and sustainable crops, while minimizing environmental impacts by reducing insecticide use and increasing utilization of beneficial insects. As climate changes, and pest spectrums and efficacy of control options change, it will be important that IPM systems evolve. As an example, in Australia higher temperatures have increased the frequency, abundance and development rate of some insects, e.g. silver leaf whitefly *Bemisia tabaci* – biotype B (Gregg and Wilson, 2008). More than ever, IPM is needed for this pest as existing controls for other pests exacerbate the problem and existing control measure are limited and expensive.
- Effective monitoring and use of predictive models to improve timing of pest management interventions. For insect management in cotton there are numerous monitoring techniques tailored for management of individual insects (Wilson *et al.*, 2004) within a cropping cycle, and many are coupled with decision support systems linked to climate (Hearn and Bange, 2002). There is also significant opportunity to further develop pest forecasting systems that can be used to predict the effects of climate change. As an example, a simulation model already used in the Australian cotton industry for *Helicoverpa* spp. is the HEAPS (HElicoverpa Armigera and Punctigera Simulation) model, which has been used to assess movement of adult moths within a regional cropping system (Fitt *et al.*, 1995). HEAPS includes modules for spatial representation in the region, moth movement, oviposition, pest and crop development and pest mortality. It will be important that decision support tools reflect current and future climate change and not capture long-term historical views.
- Effective industry and on-farm hygiene and biosecurity.
- Landscape-scale management involving groups of growers cooperating to reduce communal threats. Hoque *et al.* (2000) showed that area-wide management groups, through communicating and coordinating pest management strategies, were better able to implement insect IPM strategies that delivered economic benefits. Recent research also highlighted the need for consideration of habitat type, and spatial and temporal distribution of habitats, to ensure the potential of entomophagous arthropods to suppress economically important pests (Schellhorn *et al.*, 2014).
- Implementation of industry-wide strategies to prevent build-up of weed and insect resistance to pesticides. This includes steps to protect transgenic cultivars containing crop protection traits.

4.4. Water and Irrigation Management

Water management capabilities could play a major role in enabling producers to maintain economically viable operations as the climate changes. Irrigation is important in helping the plant mitigate the detrimental effects of high temperature. Plant capacity to moderate tissue temperature through transpirational cooling is dependent upon an adequate moisture supply. However, water use for cotton irrigation will have to compete with industrial and urban municipal use due to dwindling ground and surface water supply in many areas. Policy makers will have to decide how this limited resource will be best divided amongst the various stakeholders. There is a need to modify current crop management, which was developed assuming reasonable access to water, to management that accepts that water will always be limited. Continued investigations to implement alternative irrigation amd agronomic practices will be needed to produce profitable and high quality cotton crops consistently with less irrigation water.

Cotton production in many regions can be described as rainfed, partially, or fully irrigated. Irrigation generally serves to reduce the impacts of rainfall variability; however, future climate change in some regions may increase the risk of in-season drought or simply reduce the amount of water available for irrigation. In addition, when increases in atmospheric VPDs occur (through either reductions in humidity, increases in temperature or both), crops may still use more water despite reductions in stomatal conductance under [eCO_2].

There are a number of practices that cotton growers can use to improve WUE or to adapt to water-limited situations. These include:

- Implementing systems that monitor and assess whole-farm WUE to identify parts of the system that are inefficient. Growers consistently adopt practices to improve water storage and system irrigation efficiencies, and reduce transmission and application losses.
- Use of alternative irrigation systems such as lateral moves, centre pivot, or drip irrigation systems (Bordovsky *et al.*, 1992; Sorenson *et al.*, 2011).
- Better scheduling of irrigation utilizing technologies that continuously monitor weather (automatic weather stations), crop soil water use (capacitance probes, neutron moisture meters) and plant stress (canopy temperatures, stem diameter), but allowing for differences in soil types, demands of the crop (crop stage) and climatic conditions (e.g. temperature and humidity).
- Improving soil management by adopting controlled traffic and reduced tillage practices to minimize compaction, thereby improving soil structure and increasing the rooting zone.
- Changes in sowing time to shift periods of maximum water use into periods of lower temperatures or VPDs.
- Using reactive strategies to track climate variation on daily or seasonal time steps (Howden *et al.*, 2010). Decisions relating to irrigation management can be based on soil moisture storage, seasonal average rainfall, short- and long-term forecasts of weather and climate (rainfall and/or crop evaporative demand), and financial and commodity forecasts on a single field or whole-farm basis (Power *et al.*, 2011; Power and Cacho, 2014). At a field level, Brodrick *et al.* (2012) reported that there are opportunities to vary the timing of irrigation utilizing short-term (3 to 4 day) forecasts of evaporative demand. They demonstrated that when the soil-water deficit for irrigation is reached and when the forecast for evaporative demand is low, irrigation could be delayed without affecting yield or fibre quality. In many instances, it also increased the time for the crop to capture rainfall, reducing the need to deliver irrigation water to the crop. At a farm level, water management could be improved on a season-to-season basis if system-wide allocations were forecasted earlier (using longer range seasonal climate forecasts). This provides better insights into farm water allocation used for planning cropping areas and level of inputs (Ritchie *et al.*, 2004).
- Reducing the risk of crop failure by reducing the land area of cotton grown to increase water delivery (mega-litre per hectare) from irrigation supplies before the season begins. Determining the area to plant is a decision that has to consider the yield, and thus the water needed (accounting for climatic risk and system irrigation efficiencies) to break even (Hearn, 1992). Simulation technologies, such as those in HydroLOGIC (Richards *et al.*, 2008), which incorporates OZCOT (Hearn, 1994) that estimates yield with different water allocations and climatic impacts (including rainfall variability), can be used to assess cotton cropping areas. Recent advances in field irrigation management have included the development of a framework 'VARIwise' that develops and simulates site-specific irrigation control strategies (McCarthy *et al.*, 2010). VARIwise divides fields into spatial subunits based on databases for weather, soil and plant parameters to better account for field variability. The OZCOT model is used in two capacities in VARIwise: (i) to simulate the performance of the control strategies; and (ii) to calculate the irrigation application that achieves a desired performance objective (e.g. maximized bale yield or water productivity). These strategies can be extended to management of water and optimizing water use on a whole-farm basis.
- Better utilizing stored soil water collected from crop fallows and employ practices to capture and retain soil moisture. Strategies such as reduced tillage and stubble retention are becoming standard practice for moisture conservation. Use of rainfall to establish crops rather than pre-irrigation or 'watering-up' are worth considering, especially if there is flexibility in planting time.
- Avoiding excess nitrogen fertilizer, which encourages extra vegetative growth, subsequently lowering crop WUE. This also reduces nitrous oxide emissions (a GHG).
- Utilizing supplemental irrigation strategies or modified row configurations (e.g. skip rows) to enhance crop access to soil moisture. These strategies are not necessarily the most water use efficient, but offer significant risk mitigation in years where rainfall is limited (Montgomery and O'Halloran, 2008). These strategies also rely on reducing the length of time where the crops are in stressed conditions during the flowering period. In general, the strategy in limited water situations is to keep irrigating until irrigation water runs out.
- Considering rainfed production in those years where prices for cotton are profitable and there is a season forecast for reasonable rainfall (Bange *et al.*, 2005). Skip-row configurations can also offer significant insurance against losses in both yield and quality in those regions and years where rainfall is highly variable, and can reduce input costs such as transgenic licence fees which are priced on an area basis (Bange *et al.*, 2005). However, in areas where there are significant limitations to growth, such as substantial soil constraints or short growing seasons, it is better to either avoid skipped rows or even consider narrower row spacing to improve yield. Skip-row configurations need to be considered in the context of reducing ground cover that allow for increased soil evaporation, as well increasing the risk of runoff and soil erosion (Howden *et al.*, 2010).

- Extending the length of fallows to capture rainfall, especially on soils with a greater plant available water holding capacity.
- Shortening the time to crop maturity. To cope with limited water availability, one option would be to reduce the time to maturity and then manage a crop to achieve a targeted economic yield threshold. Crop maturity can be manipulated by choice of cultivar, insect management, nutrition, growth regulators, or late-season irrigation management (Roberts and Constable, 2003). Early crop maturity may avoid fibre quality downgrades, save water and reduce the need for late-season insect protection. However, this consideration needs to be balanced against reduced lint yield due to shorter periods of reproductive growth and maturity, as discussed previously in the section on planting time and season length (Bange and Milroy, 2004). Roberts and Constable (2003) and Bange et al. (2006) have shown that after cultivar choice, the main factor driving differences in crop maturity is fruit retention. Transgenic cultivars, which can withstand early pest damage from lepidopteron species, thus maintaining higher fruit retention and avoiding plant terminal damage, can achieve similar yields to non-transgenic cultivars and use less water by maturing earlier (Richards et al., 2006).
- Exploring the use of degradable polymer films as mulches in cotton systems, such as those described in Braunack et al. (2015). One major issue with using plastic film as mulch has been the problem of disposal because it did not degrade (Shogren, 2001). However, new formulations that degrade to water and carbon dioxide (oxodegradable films) overcome this limitation and avoid plastic mulch disposal in land-fill (Dai and Dong, 2014). Thin plastic films have been used to increase soil temperature, conserve soil water and to improve crop establishment for cotton (Stathakos et al., 2006; Dai and Dong, 2014). There are advantages in both rainfed and irrigated situations by conserving available water and improving crop WUE (Wang et al., 2009; Zhou et al., 2012; Bu et al., 2013; Zhang et al., 2013). It may also be possible to harvest rainwater from the film-covered areas (Ruidisch et al., 2013; Liu et al., 2014). However, consideration would need to be given to field layout for runoff and erosion potential due to slope.
- Reducing the risk of waterlogging. This can be achieved through appropriate field design to ensure adequate drainage and runoff, growing cotton on well-formed hills, and avoiding irrigation before significant rainfall events by monitoring weather forecasts. The use of nitrogen and iron foliar fertilizer applied prior to waterlogging may have beneficial effects on yield (Hodgson and Macleod, 1987). Similarly, the use of the growth regulator AVG applied prior to waterlogging may have beneficial effects by maintaining photosynthesis, improving node production and reducing fruit abscission (Bange et al., 2010c; Najeeb et al., 2015b). Waterlogging suppresses cotton growth and yield by reducing leaf number, inhibiting photosynthetic processes, and increasing expression of genes regulating synthesis of the ethylene precursor 1-aminocyclopropane-1-carboxylate (ACC) (Christianson et al., 2010; Yang et al., 2011). Synthesis of ACC and its concomitant conversion into ethylene can be controlled by application of the chemicals AVG and aminoethoxycetic acid and cobalt (Co^{2+}) that inhibit the activity of the enzyme ACC-oxidase of ethylene biosynthesis pathway (Chaves and Mello-Farias, 2006). The use of other plant growth regulators (PGRs) (Cothren, 1995) also offers opportunities to support management practices to reduce impacts of other abiotic impacts generated by climate change, such as heat and water stress.

4.5. Management of Cotton Crops with Plant Growth Regulators

Plant growth regulators (PGRs) are commonly used in cotton production systems in order to control and manipulate growth, which is mainly regulated by endogenous plant hormones. Global climate change is expected to affect plant growth and PGRs may be an important management tool to ensure optimal and sustainable yields.

As mentioned previously, increases in season length are anticipated due to global warming, and this could result in higher yields. However, research has indicated that with longer seasons, vegetative growth needs to be prolonged until early boll set to allow crops to 'cutout' and mature later in order to achieve higher yields (Yeates et al., 2010; Constable and Bange, 2015). PGRs, such as mepiquat chloride (1,1dimethylpiperidinium), can enable farmers to control plant growth and achieve maximum results (Krieg and Kerby, 1985; Constable, 1995; Reddy, V.R. et al., 1995; Reddy, A.R. et al., 1996). In some regions, growing crops with high inputs, [eCO_2] and with warmer temperatures early in the season, may predispose crops to excessive vegetative growth. Without canopy control using mepiquat chloride, excessive vegetative growth may reduce yields and increase early water use. In addition under conditions of high temperature, Rosolem et al. (2013) showed that the rate of mepiquat chloride to be applied must be disproportionately increased, because either the plant is impaired by high temperature lessening the effect of mepiquat, or degradation of the mepiquat in the plant is more rapid.

Increased frequency of temperature extremes is predicted (and currently occurring) as a result of climate change, with detrimental effects on plant physiological processes and yield outputs. Reddy et al. (1991b) suggested that one of the main losses in reproductive capacity in cotton with elevated temperatures associated with climate change was increased fruit abscission. Enhanced ethylene synthesis is associated with a wide range of injuries and stresses in cotton and is responsible for young fruit abscission.

Similar to AVG, 1-methylcyclopropene (1-MCP) is a PGR and is commonly applied as an anti-ethylene growth regulator to many horticultural crops to reduce the harmful effects of ethylene. It arrests ethylene synthesis by blocking its receptor sites in plant cells and controls ethylene-induced cell membrane electrolyte leakage (Blankenship and Dole, 2003). Applications of 1-MCP at first flower and 2 weeks after flowering significantly increased seed cotton and lint yield compared to untreated plants (Kawakami et al., 2010); the positive effect was greater under conditions of heat stress (Storch and Oosterhuis, 2009).

Djanaguiraman et al. (2010) reported that foliar application of Chaperone (a PGR containing nitrophenolates, sodium 5-nitroguaiacolate, sodium ortho-nitrophenolate and sodium para-nitrophenolate, that enhances plant protein formation) reduced boll abscission and resulted in higher yields by alleviating reactive oxygen species effects through an increase in antioxidant capacity. Chaperone increased nitrogen and protein content of floral components in cotton (Oosterhuis and Brown, 2005), particularly under high temperature conditions, and thereby enhanced reproductive growth and yield.

Polyamines (putrescine (PUT), spermidine (SPD) and spermine (SPM)) are another group of PGRs that are used in horticultural crops, but not commercially available for row crops. Their indispensable presence at flowering and early fruit development, and antagonism to ABA, make them a potentially powerful growth regulator. Application of PUT to cotton 24 h prior to flower opening increased seed set and alleviated the detrimental effects of increased temperatures (Bibi et al., 2010a). However, the consistent effect of these growth regulators in cotton has not yet been demonstrated.

Water-deficit stress events are expected to increase in frequency and intensity due to global warming. Zhao and Oosterhuis (1997) reported that PGR-IV (a combination of gibberellic acid, indolebutyric acid and a propriety fermentation broth) was able to partially alleviate the effects of water stress on photosynthesis and dry matter accumulation; however, experiments were conducted with potted plants in growth chambers and not under field conditions. Similar research with 1-MCP application on water-stressed cotton plants indicated that 1-MCP was unable to protect against decreases in leaf photosynthesis and stomatal conductance (Loka and Oosterhuis, 2013).

Seed pre-treatment with PGRs is commonly practised for improving tolerance to abiotic stresses including drought, waterlogging and salinity (Liu et al., 2012). Application of plant growth hormones such as BAP (6-benzylaminopurine, benzyl adenine is a first-generation synthetic cytokinin) and GA_3 (gibberellic acid) alleviated waterlogging and drought stress in cotton by maintaining turgor pressure and blocking stomatal closure (Goswami, 1990). Foliar application of 5 µM ABA (abscisic acid), IAA (auxin-indole-3-acetic acid) and BAP just before waterlogging stress increased pyruvate content and photosynthetic metabolites of cotton under waterlogged condition (Pandey et al., 2001).

Obviously, PGRs can play a role under global climate change in controlling growth and protecting yield, but research is needed to adapt these growth regulators for particular farming systems and climate change scenarios.

4.6. Crop Diversification with Crop Rotations and Cover Crops

Farming has always fundamentally been about effectively optimizing and responsibly managing natural resources for generating livelihoods. In future climates, growers may need to be more flexible with their cropping choices to ensure that the farming system is viable and resilient. Commodity prices of other crops (such as wheat, maize and sorghum) can provide attractive options for generating returns similar to cotton production. Growers can utilize and substitute their farm resources (including water) for growing other crops in rotation, or as cover crops. Cotton may become part of a more variable and viable crop rotation programme.

Wheat and maize in rotation with cotton have been shown to raise cotton yields due to improvements in soil physical structure (Hulugalle et al., 2005b, 2007) and reductions in disease outbreaks. Cover crops can also be grown to reduce long fallow periods in a cropping cycle specifically to protect the soil from erosion and reduce nutrient loss through erosion or leaching. Once the cover crop is killed prior to planting of the cash crop, residues of the cover crop that remain on the soil surface protect the soil until canopy closure of the cash crop. In terms of protecting crop production from the effects of climate change, soil productivity is enhanced with the use of cover crops by reducing soil erosion, increasing soil organic matter, recycling nutrients, moderating soil temperature and, in the case of legumes, adding nitrogen to the soil (Unger and Vigil, 1998). In general, cover cropping has not been commonly used in modern cotton production systems. In the south-east USA, cover-crop use has increased slightly, as in the 1990s farmers started to adopt conservation tillage systems because cover crops tend to improve cotton productivity in conservation systems (Bauer and Busscher, 1996; Raper *et al.*, 2000). Diversified farming practices, such as incorporating cover crops as part of cotton rotations, are difficult to incorporate into highly capitalized mechanized systems (Pearson, 1967; Rochester and Peoples, 2005). There are some specific considerations that relate to the use of cover crops:

- Increasing total water consumption for cotton production by growing cover crops to improve soil will increase the water footprint (Hoekstra and Hung, 2002) of cotton products. There is a need to quantify when and where (or if) the increase in available soil water to cotton through soil improvements by cover crops maintains total water use per kilogram of lint similar to current ratios. Establishing water use requirements of species used as cover crops will be a first step towards attaining this goal.
- Incorporation of cover crops into cotton farming systems involves a trade-off between water used by the cover crop and water that could be kept in the soil to support the cotton crop. Unger and Vigil (1998) suggested that cover crops are often suitable for humid and sub-humid regions, but not for rainfed cropping systems in semi-arid regions because depletion of soil water by the cover crop causes decreased yield of the succeeding crop. They suggested that for semi-arid regions, use of conservation management with the crop residues remaining on the surface is preferable to growing cover crops because it will provide many of the benefits of growing cover crops except for nitrogen fixation, prevention of nutrient leaching and additional organic inputs.
- Reducing cost of establishing cover crops. Costs associated with establishing cover crops are a major deterrent to expanding their use. Using hard-seeded cover crop species that can naturally reseed prior to cotton planting is attractive because it reduces cost of establishment and management. Efforts need to be made in genetic screening and enhancement of species capable of reseeding prior to cotton planting in the spring. For non-reseeding annuals, efforts are needed to increase the reliability of broadcast seeding of cereal and other species into cotton both prior to and after defoliation.
- Planting species-diverse cover crop mixtures. Cover crop mixtures that contain a mixture of plants differing in ecosystem enhancing traits, such as N-fixation, rooting depth and distribution, growth rate and leaf area index, may increase biomass potential (Wortman *et al.*, 2012) and have not been widely evaluated for cotton management. Planting diverse mixtures of cover crops with seeds that vary in size, density, weight and seeding rate will require innovative engineering advances in seeding equipment so that these can be used in large farming systems.
- Increasing the opportunities to raise the economic value of cover crops for uses like grazing (Schomberg *et al.*, 2014) or for potential feedstocks for biofuel (Raper *et al.*, 2011).
- Managing spatial variability of cotton grown following cover crops. The use of precision agriculture technologies such as Normalized Difference Vegetation Index (NDVI) could help to resolve problems of variable nutrition prior to the cotton production cycle.

It has been suggested that coupling both rotation and cover crops with conservation tillage management will reduce the effects of climate change on crop production (Lal *et al.*, 2011; Walthall *et al.*, 2012). Growing rotation or cover crops during cotton fallow periods likely will increase soil C sequestration, thereby reducing the amount of CO_2 in the atmosphere (Rochester and Peoples, 2005). Growing legume cover crops as a source of N will also reduce the amount of fertilizer-N applied to the fields, and reduce CO_2 emissions generated in the manufacture of fertilizer N (Rochester *et al.*, 2001a, b). Legume-based cropping systems may also produce less atmospheric nitrous oxide emissions from fields than when N fertilizer is applied, although there is risk of nitrous oxide emissions from green manure legumes (Jensen *et al.*, 2012). Contrary to many other practices used to produce cotton, use of rotation and cover crops may not result in economic gain in the short term. Research is needed that will lead to reduced economic risk associated with implementation of rotation and cover crops so that they will be incorporated in large-scale cotton operations to achieve long-term soil improvement (Reeves, 1994).

4.7. Utilizing Seasonal Climate Forecasts

Cotton production systems are dynamic and responsive to fluctuations in climate. Risk, or the chance of incurring a financial or environmental loss, is a key factor influencing decision-making

(Hardaker et al., 1997). Coupled with on-the-ground observations of a system (e.g. soil moisture), the skill in seasonal climate forecasting offers considerable opportunities to managers via its potential to realize system improvements (i.e. increased profits and/or reduced risks).

The use of seasonal climate forecasts, such as the El Niño-Southern Oscillation system (ENSO), has provided a basis for taking advantage of climate variability, rather than passively accepting the risk it generates (Hammer and Nicholls, 1996). Adjusting management to accommodate predictions of climate during the upcoming growing season offers considerable opportunities to managers of agricultural systems. In Australia, cropping management strategies have been assessed by the use of crop simulation technologies employed with assessments of historical climate records partitioned into groups based on the Southern Oscillation Index (SOI) (Stone and Auliciems, 1992). The most successful management options are then assessed in terms of both their average and risk.

In cotton systems, the SOI approach has been used to develop management strategies, including selection of skip-row rainfed configurations (Hammer, 2000), choice of crop rotation (cotton/sorghum versus fallow) (Carberry et al., 2000) and cropping area (Ritchie et al., 2004; Power et al., 2011). Hammer (2000) was able to demonstrate that system economic performance could be improved when responding tactically each season to the forecast (an 11% increase in profit over the complete historical record). Essentially, when drier years were forecasted, wider skip configurations were chosen to reduce the impact of less rainfall on yield. Similarly, Carberry et al. (2000) showed that when summer rainfall was forecasted as being favourable, a cotton or sorghum crop could be included to increase system profit compared to a summer fallow without any significant increase in risk. Bange et al. (1999) also showed in irrigated cotton systems that there was a relationship of forecasted wetter seasons associated with lower yield performance compared with the average.

These studies demonstrated that tactical responses to a forecast may pay off on average, over a period of years, but there can be no guarantee for the ensuing season. Therefore, effective implementation of seasonal forecasts requires understanding the risks associated with the forecasts. Forecasting skill is imperfect and approaches to applying the existing skill to management issues are still evolving. Additionally, the impact of climate change on seasonal forecasts such as ENSO remains uncertain (Howden et al., 2010).

4.8. Optimizing Efficiency of Resource Inputs

Improving production efficiencies can improve crop returns and allow more resources (e.g. water from improved WUE) to be accessed for further production. Growers are faced with rising costs of fuel, fertilizer and chemicals, future costs associated with 'carbon pollution', environmental concerns, and possible reductions in available irrigation water (from government intervention in water regulation and climate change), which will all impact cotton productivity. Therefore, improvements in production efficiencies need to be part of the strategies to cope with future climate change. Two important and key foci in cotton production systems in a changing climate are the efficient use of nitrogen fertilizer and water.

4.8.1. Crop nitrogen use

Monitoring soil fertility and crop nutrient uptake are important because growers realize the importance of avoiding nutrient deficiency and the expense and environmental concerns (including GHG emissions) associated with excess fertilizer use. Excess N fertilizer can also lower WUE or yield by encouraging excessive vegetative growth and delaying maturity. Decision support systems (Rochester et al., 2001a) are used by growers to provide information for determining the appropriate rates for N fertilizer use and the need for other nutrients, based on crop stage (utilizing climate information) and performance. Since the effects of rising atmospheric [CO_2], warmer temperatures and changing rainfall on plant growth can be partially modified by nutrient availability, it will be important to monitor and adapt fertilizer practices with climate change. Indeed, monitoring approaches such as leaf and petiole tests will also need constant revision since plant C:N ratios are often lower in cotton grown in [eCO_2] (Wu et al., 2007).

Current estimates of N requirements of high yielding crops in Australia are in the range of 240–270 kg N ha^{-1} crop uptake (Rochester and Constable, 2015). A survey of Australian cotton fields by Rochester (2007) highlighted that a significant proportion (17 in 34 cotton crops) had low N use efficiency (kg lint per kg N uptake) as a consequence of excessive N fertilizer application. They calculated on average an excess of 40 kg N ha^{-1} was applied, increasing the chances of N being lost from the system and contributing to GHGs. A major source of anthropogenic N_2O emission is the use of N fertilizers in agriculture. As a substantial proportion of applied fertilizers are emitted in the form of N_2O, better targeted fertilizer applications, which reduce the availability of N to microorganisms, can substantially decrease N_2O emissions (Singh et al., 2010). In cotton farming, it is estimated that up to 1% of N fertilizer applied to an irrigated cotton crop is lost as N_2O gas (Rochester, 2003), while Maraseni et al. (2010) compared three cotton farming practices (irrigated-solid plant, rainfed-solid plant and rainfed-double skip) and found that irrigated cotton farming emits greater than 12 times more GHG per hectare than from rainfed cotton farming.

While optimizing fertilizer use to a particular production system is a clear strategy for reducing losses, strategies such as the use of crop rotations and cover crops (already discussed) and nitrification inhibitors and biochar may reduce losses. One approach for crop N nutrition is to supplement or even replace entirely the use of artificial N fertilizer with nitrogen fixed by legumes. Cotton crops can be grown with N provided entirely by legumes (Rochester et al., 2001b), with vetch (Vicia villo) especially able to supply high-yielding crops. The additional benefits of utilizing legumes in the cropping system include improvements in soil structure and soil health. The use of nitrification inhibitors in fertilizers to limit nitrate production and subsequent leaching or denitrification losses is now a well-established strategy (Smith et al., 2008; Singh et al., 2010; Kawakami et al., 2012). Rochester et al. (1996) examined the efficacy of nitrification inhibitors to reduce N loss through denitrification in flood-irrigated cotton field soil. They found that etridiazole was the most effective at reducing N loss, most likely by suppressing the production of nitrate rather than inhibiting the consumption of nitrate by denitrifiers.

4.8.2. Crop water use

Tennakoon and Milroy (2003) and Roth et al. (2013), in their reviews of cotton WUEs, highlighted that there were significant opportunities to improve WUE at all levels (whole farm to agronomic). Their analyses showed that irrigated cotton farms had significant losses through conveyance, storage and application losses, or improper scheduling. Overall, there continues to be a need for crop management and research efforts to remain diligent in challenging the paradigm of cotton systems that were developed assuming reasonable access to water, to one that accepts that water will always be limited. A range of current options already discussed with alternative irrigation and agronomic practices that maximize crop husbandry and water use will be needed that can consistently produce profitable and high-quality cotton crops with less irrigation water.

While improvements in leaf level transpiration efficiency of cotton have been identified with [eCO_2], there is little current evidence that suggests that there will be improvements in WUE at the whole-canopy level as water use increases. Reddy, K.R. et al. (1996) in the SPAR facility found that there were no changes in water use as the increase in early canopy leaf area development from improved photosynthesis negated the benefits of improved leaf level transpiration with [eCO_2] (700 µmol mol^{-1}). Studies conducted by Samarakoon and Gifford (1996) in controlled environments also found that cotton growth, leaf transpiration and WUE was improved when [CO_2] was increased to 710 µmol mol^{-1}. Thus leaf-level water savings occurred concurrently with substantial increases in growth and leaf area, thereby resulting in water use per plant being 40–50% higher than in ambient [CO_2]. They also found that in drying weather conditions, this resulted in soil drying faster under high CO_2 conditions, accelerating the onset of water stress compared with the ambient CO_2 treatment. This experiment was conducted in pots in a glasshouse. Effects of water stress on growth may be exhibited earlier in pots compared with field studies (Jordon and Ritchie, 1971), so additional work needs to be conducted in field environments.

In [eCO_2] (550 µmol mol^{-1}), Hunsaker et al. (1994) reported in their FACE experiments that there were no changes in seasonal crop evapotranspiration in both well-watered and water-stressed treatments. Yield was improved in [eCO_2], which led to improvements in agronomic WUE (kg lint mm^{-1} Et) (Mauney et al., 1994). Improvements in WUE were attributed to improved growth associated with increased leaf area (Mauney et al., 1994) and root function (Prior et al., 1994) (both direct effects) and a longer flowering period (an indirect effect).

The contrasting results of the effects of [CO_2] on cotton evapotranspiration and subsequent water use needs resolution especially if cotton is subjected to the additional confounding effects of limited water availability, and increased atmospheric evaporative demand due to changes in both temperature and humidity. Higher evaporative demand in well-watered crops also has the potential to increase transpiration and lower WUE (Rawson and Begg, 1977), and substantially hotter temperatures may potentially negate benefits of [eCO_2] by increasing transpiration and reducing leaf photosynthesis (Reddy, K.R. et al., 1996).

If water is limited, increased water use early in the cotton season as a function of [eCO$_2$], warmer temperatures and high inputs of nutrition may reduce water availability for later reproductive growth. In particular, this may be true due to increased vegetative growth in high-input/high-yielding systems. While the FACE experiments (Hunsaker et al., 1994) did not show changes in water use over the season, the yield levels and inputs in these experiments were lower than commonly observed in current farming practices. Mauney et al. (1994) reported that the lint yields of the cotton FACE experiments conducted in Arizona (USA) were in the vicinity of 1500 to 1600 kg ha^{-1} (actual yields not reported) and N inputs were 130 kg ha^{-1}. Constable and Bange (2015) recently reported yields in cotton systems (ambient [CO$_2$]) in excess of 3000 kg ha^{-1} and N rates approaching 300 kg ha^{-1}. Practices restricting early vegetative growth to limit water use early in the season may be necessary if vegetative growth is greatly increased due to [eCO$_2$] fertilization. Management to restrict early vegetative growth (resulting in season-long reductions in transpiration) may include the use of growth regulators or cultivars that develop fruit early to act as a sink restricting vegetative growth.

4.9. Soil Management

Global climate change projections include higher temperature and a change in rainfall patterns for many cotton-growing areas. Whether the change in rainfall patterns results in reductions in total precipitation or just changes in seasonal distribution, management practices are needed to reduce water-deficit stress. Water use efficiency can be increased through the incorporation of soil management techniques that increase infiltration and reduce soil water evaporation.

Periods of drought are predicted to occur more often in some regions as a result of global climate change. If this occurs, growers will need farming practices that increase the crop water availability from precipitation. One widely used practice that may help mitigate the effects of prolonged rain-free periods is conservation tillage. Conservation tillage can increase soil water, especially before the soil is covered with the crop canopy, and can increase crop yields compared to conventional tillage (Phillips et al., 1980). Higher yields with conservation tillage have been attributed to reduced evaporation from the soil surface (Lascano et al., 1994) and reduced runoff (Truman et al., 2003).

Crop residues have been suggested to be an important component of conservation tillage production systems (Langdale et al., 1990). The amount of residue needed for improving soil productivity in the south-east USA has been estimated to be more than 12 Mg ha^{-1} (Bruce et al., 1995). In conservation tillage systems for cotton, crop residues left on the soil surface following cotton harvest are typically minimal. Previous research indicates that cover crops (Bauer and Busscher, 1996; Raper et al., 2000) or rotations with high-residue crops (Bordovsky et al., 1994; Reddy et al., 2006) have the potential to increase the productivity of cotton in conservation production systems. Several studies conducted by Hulugalle et al. (2005b) found that conventional tillage regimes often resulted in lower soil organic C than minimum tillage regimes.

It has been suggested that soil C storage is unlikely to be modified by changes in crop rotation practices for agricultural systems in semi-arid zones (Hulugalle et al., 2013), as crop rotation only influenced soil organic C concentrations of the surface soil and had a negligible impact on the subsoil. However, on a different soil type with less sodicity, Rochester (2011) demonstrated that soil C could be increased over time with minimal incorporation of crop residues in crops producing large amounts of biomass in both the cotton and rotation phases.

A study by Kaisermann et al., 2013 found that agricultural management (i.e. of tillage and fertilizer) can increase microbial functional resistance to drought stress via establishment of bacterial communities with particular metabolic capacities. While large losses of soil organic C owing to cultivation also can be reduced by low- and no-tillage practices, this can, however, favour soil communities dominated by fungi (Castro et al., 2010). Such agroecosystems prevent the increase in microbial decomposition and respiration that comes from soil disturbance (Smith, 2008) and in some cases can enhance emissions of N$_2$O from soil (owing to increased rates of denitrification that are due to anaerobic conditions in compacted soils). This offsets some of the benefits of increased soil carbon storage (Singh et al., 2010).

Another emerging concern related to modern cotton production systems is the loss of soil health and resilience that pertains to soil compaction caused by machine cotton pickers that have on-board module-building capabilities. These pickers are being used to reduce costs relating to picking and introduce labour savings (van der Sluijs et al., 2015); however, they are heavy (32 t) and have the potential to increase compaction in the subsoil, limiting efficiencies in both water and nutrition (Braunack and Johnston, 2014). Future management of soils utilizing these picking systems will need to consider strategies to: (i) ameliorate compaction using crop rotations that dry the soil profile; (ii) further implement controlled traffic systems; and (iii) seek to reduce moisture in the profile at picking. In the case of irrigated systems, it will be necessary to have appropriate scheduling of the last irrigation to reduce the risk of moist profiles at the end of the season.

V ROLE OF RESEARCH IN MODERN COTTON SYSTEMS ADAPTING TO CLIMATE CHANGE

The cotton industry covers a large geographical region and thus is already experiencing a wide range of climatic extremes. Subsequently, technologies and systems have been developed to mitigate high temperature and water stress. Photosynthetic acclimation occurs in cotton (Downton and Slatyer, 1972) and plants occupying thermally contrasting environments generally exhibit photosynthetic responses that reflect adaptation to the temperature regimes of their respective habitats (Berry and Bjorkman, 1980). For example, cotton is successfully grown at temperatures in excess of 40°C in India and Pakistan (e.g. Table 5) indicating some adaptation and successful breeding selection.

In the face of climate change, along with market competition from synthetics and price volatility, cotton-growing enterprises will need to continue to produce higher yields, improve resource use efficiency and be more resilient to maintain profitability.

Key approaches to raising and maintaining yields are to develop and refine new technologies (e.g. precision agriculture, cultivars with both yield and fibre quality improvements, chemicals etc.), agronomic practices (e.g. sowing time, plant population, crop nutrition etc.) and management systems (e.g. IPM, integrated weed management etc.) that enable cotton to grow healthier or more tolerant to both abiotic (temperature and water stress, waterlogging) and biotic stresses (pests and diseases). Overall, detailed integrative research over a greater range of environments and stresses is needed to properly assess impacts and adaptation options that translate into realized yield and quality improvements (Sankaranarayanan et al., 2010). Few studies have been conducted in any crop that deal with three-way interactions of changes in combinations of [CO_2], water, temperature and atmospheric humidity (Jagadish et al., 2014). Future cotton research programmes will need to ensure that knowledge and technologies respond to these impacts and strategies are developed to both exploit and avoid maladaptation to climate change.

In a recent review by Hatfield and Walthall (2015), an emphasis was placed on leveraging opportunities by adopting a genetic × environment × management (G×E×M) interaction as a foundation approach to meet global agriculture needs and realizing potential of cropping systems in future climates. This approach will be needed to meet the challenges faced by changes in cotton-cropping systems. There are few studies in cotton that have demonstrated the value of G×E×M to improve cotton productivity. Analyses by Liu et al. (2013), using their advanced line trials containing varieties grown over a 30-year period from 1982 to 2009, demonstrated that yield gain in the Australian cotton industry resulted from improvement in varieties (G; 50% improvement), in crop management (M; 26% improvement) and from the interaction between improved varieties and improved management (G×M; 24% improvement). The challenge remains on how to exploit the G×E×M interaction in research and deliver the benefits to cotton growers. Synthesizing the review of Hatfield and Walthall (2015) they contend that the following ingredients in research approaches to G×E×M need consideration to successfully meet the future challenge. Many of these have been highlighted during the course of this review:

- A focus on soil improvement to remove factors for water and nutrient accessibility.
- An incorporation of multidisciplinary science in research teams.
- Development of robust tools to assess photosynthetic efficiency (at the leaf and canopy level).
- An understanding of why crops are not achieving their potential (at the regional and local level).

Table 5. Comparison of the highest monthly average maximum and minimum temperatures (degrees Celsius) for Narrabri (New South Wales, Australia, source: Australian Bureau of Meteorology), Maricopa (Arizona, USA) and Multan (Pakistan) (source: http://www.weatherbase.com) during their respective summer production seasons. Yield potential is significantly less in Maricopa and Multan as a result of the hot temperatures.

Location	Summer maximum	Summer minimum
Multan, Pakistan	42.3 (Jun)	28.7 (Jul)
Maricopa, Arizona, USA	41.6 (Jul)	24.1 (Jul)
Narrabri, New South Wales, Australia	33.8 (Jan)	21.4 (Jan)

- Adoption of innovative technologies as part of the G×E×M approach (e.g. precision agriculture).
- Involve the grower in applied research to improve outcomes and uptake.
- Characterize plant and crop responses to stress and develop rapid screening and monitoring systems.
- Utilize crop simulation models to assess potential alternative scenarios for differing germplasm with different management to ascertain opportunities in the G×E×M approach as well as understanding how future climate affects these outcomes.

In this section, we summarize key research considerations needed to meet these challenges. We will discuss these research needs under the headings of research into genetic improvement and cotton physiology, soil management, and management of the cotton crop in a whole systems context.

5.1. Genetic Improvement and Cotton Physiology

Cultivar choice is a strong component of realizing both target yield and fibre quality levels on a cotton farm. A delicate balance needs to be resolved between yield, fibre quality, price and other important considerations such as disease resistance, and insect and herbicide resistance. Developing improved understanding of the physiology and the genetics underpinning responses of cotton genotypes to abiotic stress offer substantial opportunities for regional cotton industries to deal with many elements of predicted climate change.

For cotton breeders, delivering commercially available high-yielding cultivars to cotton growers remains a necessity, such that cotton systems remain economically viable. High selection pressure on yield remains a successful means to capture tolerance to both abiotic and biotic stress. Amongst records of improving yields across regions there is also evidence that this approach has been successful in generating tolerance in high-yielding cotton cultivars for abiotic stress. Specific tolerances for heat (Constable *et al.*, 2001; Bibi *et al.*, 2008a; Cottee *et al.*, 2010) and water stress in rainfed environments (Stiller *et al.*, 2005) have been recorded despite no specific selection pressure on these stresses. Genetic variability of transpiration responses to vapour pressure deficits (VPD) have been established for soybean (*Glycine max* (L.) Merr.) (Sadok and Sinclair, 2009), and it was found that a particular genotype could limit transpiration rate at high VPD and approach wilting point slowly, when under water stress. This warrants more attention in cotton systems that are water limited.

Opportunities to continue to improve yield remain possible given indications in the review by Constable and Bange (2015). They identified a number of opportunities for research to address yield potential and theoretical yield in cotton. In breeding, options for longer season and more indeterminate growth habit are required with relatively slow crop setting, but with greater final fruit numbers (Hearn, 1976a). The cotton growth habit is complex, so crop physiology studies are required to predict more accurately fruiting dynamics under high yielding conditions. A challenge for molecular biology is to increase photosynthetic capacity (e.g. Maurino and Weber, 2013; McGrath and Long, 2014) and translate this into improved canopy radiation use efficiency (RUE). Such research is still in its infancy and there are many obstacles to overcome, but there may be long-term benefits in increasing rates of photosynthesis. It was calculated that resource requirements for higher yields are more limited by nutrient uptake and distribution than by water requirements, so research in crop management is also required in understanding improved nutrient use efficiency through better uptake of soil or fertilizer nutrients and better redistribution to fruit.

Recognizing the ever increasing need to maintain yields in a variable and changing climate, many cotton genetics programmes are trying to develop germplasm with tolerance to various abiotic stresses (Allen and Aleman, 2011). Efforts have been put forth using molecular markers to identify and characterize quantitative trait loci (QTL) associated with abiotic stress tolerance in cotton (Paterson *et al.*, 2003; Saranga *et al.*, 2004). For the most part, these efforts have focused on mining traits and genetic variability within the cotton germplasm pool (*G. hirsutum* and *G. barbadense*) and other closely related *Gossypium* species. Alternatively, genes associated with targeted biochemical pathways involved in conveying a stress tolerance that come from a completely different source could be introduced into the cotton genome through transgenic technologies (Chakravarthy *et al.*, 2014). Although the use of transgenic technology can provide a more focused approach to the genetic manipulations, it also comes with its own set of problems, such as whether the inserted foreign DNA might affect native physiological processes. Nevertheless, many private and public breeding programmes are devoting resources to select for drought and temperature stress tolerance. However, these traits are highly complex, which dictate that progress will be slow. Most of the initial screening and selecting of lines has occurred in controlled environments, such as glasshouse or growth chambers. Field testing and confirmation of these stress tolerance traits has not proceeded as fast. It may still be years before stress-tolerant cultivars are available in the market. In addition, the current costs of bringing these traits to the market through strict regulatory processes make this a more difficult realization.

Of all the projected changes in climate, the ongoing rise in atmospheric $[CO_2]$ is the best documented and forecasted climate variable known to impact plant growth. In fact, the positive effects of rising $[CO_2]$ on C_3 plant metabolism may be considered a 'silver lining' around the clouds of gloomy projections involving adverse changes in climate, including higher temperatures, drought and extreme weather events. Atmospheric $[CO_2]$ has naturally fluctuated over geologic time and has been considerably lower in the past than the present. The progenitors of our modern cotton cultivars evolved during these periods of low atmospheric $[CO_2]$.

There is evidence that evolutionary adaptations to low $[CO_2]$ may limit plant responses to current and future $[CO_2]$ (Tissue *et al.*, 1995; Sage and Coleman, 2001). As noted by Ziska *et al.* (2001), $[eCO_2]$ often results in greater increases in vegetative biomass compared with reproductive biomass yield such that harvest index is reduced with $[eCO_2]$ in soybean (Baker *et al.*, 1989; Ziska *et al.*, 1998), rice (Moya *et al.*, 1998) and wheat (Manderscheid and Weigel, 1997). These results may indicate a lack of genetic optimization of current crop cultivars to the ongoing rise in atmospheric $[CO_2]$.

Because $[eCO_2]$ has been shown to ameliorate many of the negative impacts of a wide range of environmental stresses on plants, a prudent course of action would seem to be breeding or selecting crop cultivars for increased response to future $[eCO_2]$. Citing a lack of further photosynthetic or grain yield response of the rice cv. IR-30 to $[CO_2]$ above 500 ppm, Baker *et al.* (1990) suggested selecting or screening rice varieties for increased response to CO_2 in order to more fully take advantage of future increases in global atmospheric $[CO_2]$. This course of action appears to be promising for cotton in part because of the wide range of responsiveness among current crop cultivars to $[eCO_2]$ already demonstrated in other crop species including rice (Baker, 2004; Baker and Allen, 2005), soybean (Ziska *et al.*, 2001) and wheat (Manderscheid and Weigel, 1997). Recent investigations by Broughton (2015) comparing an older cultivar with a current domesticated cultivar did not show any interaction with $[eCO_2]$ (640 μmol mol^{-1}) in leaf photosynthesis, therefore highlighting an opportunity for breeding. A related question for future research becomes 'Is the often observed photosynthetic acclimation to $[eCO_2]$ an indication of a lack of genetic fitness to current and future $[CO_2]$?'

The occurrence (or non-occurrence) of 'photosynthetic acclimation', or 'down-regulation of photosynthesis', in response to CO_2 enrichment is a naturally occurring phenomenon that, due to a lack of understanding, limits our ability to predict plant and ecosystem responses to CO_2 enrichment. In some plant species, photosynthesis was initially stimulated by CO_2 enrichment but then subsequently declined with continued CO_2 enrichment (Tissue and Oechel, 1987; Sage, 1994; Drake *et al.*, 1997). Long-term CO_2 enrichment can result in decreases in the content of Rubisco and other photosynthetic pigments as well as decreases in Rubisco activity (Besford *et al.*, 1990; Rowland-Bamford *et al.*, 1991; Bowes, 1993; Tissue *et al.*, 1993). Overall, there is an opportunity to examine the effect of elevated atmospheric $[CO_2]$ and warmer temperatures on wild, old and modern cotton cultivars. It is unknown whether different cotton cultivars respond differently to these environments. Findings from this research may aid physiological trait selection in the development of cultivars for future environments.

Climate change also has the potential to negatively affect cotton fibre quality (Lou *et al.*, unpublished), through changes in limited access to water during boll filling (fibre length reductions) and increases in temperature (micronaire increases). To ensure that cotton remains an attractive fibre for use in textile manufacturing and delivers sustainable prices for growers, there remains an imperative to consider quality. Cotton spinners require longer, stronger, finer, more uniform and cleaner cotton to reduce waste, which will allow more rapid spinning to reduce production costs and to allow better fabric and garment manufacture and enhance cotton's competiveness with synthetic fibres.

A substantial challenge to cotton breeders in improving yield concurrently with fibre quality is the negative association between yield and fibre quality, which prevents the highest yielding cultivars from possessing premium fibre quality (Clement *et al.*, 2012). Current research efforts are attempting to break this association utilizing early generation selection strategies that employ the use of a yarn quality index to integrate the fibre properties of length, strength and fineness together with yield (Clement *et al.*, 2015). Genetic engineering will also potentially play a role in assisting improvements in quality and generating novel fibre traits (e.g. elongation and moisture absorption) (Mansoor and Paterson, 2012; Chakravarthy *et al.*, 2014). However, to fully realize the benefits of improving fibre quality, all cotton industries will need to work together to address the challenges and opportunities for improving quality. The task for cotton growers and industries is to optimize fibre quality in all steps from strategic farm plans, cultivar choice, crop management, harvesting and ginning. Constable and Bange (2008) have termed this 'Integrated Fibre Management' (IFM) to emphasize the importance of a balanced approach to managing fibre quality, to be analogous with approaches such as IPM. New technologies, new instruments, new decision support programmes and communication will facilitate IFM and current research in the Australian industry is investigating means to improve quality. It is also recognized that there are opportunities to improve the value of cotton as a food and fibre crop by improving the quality of cotton seed oil by removing toxic gossypol (Palle *et al.*, 2013) and altering the fatty acid composition (Liu *et al.*, 2002).

Introgressing these traits quickly and efficiently into commercial breeding material remains a significant challenge for breeders, especially in light of the need for constant yield improvements and inclusion of transgenic crop protection traits. Along with traditional approaches to breeding, future breeding efforts will also need to rely on both improved genotyping and phenotyping approaches for trait selection. To achieve this effectively and efficiently, concurrent investments will be needed in: (i) identifying the genes related to the traits of interest; (ii) developing a understanding of the physiological responses that changes in the genes affect; and (iii) cost effective means of phenotyping the plants/crop response such as those being assessed in cotton by Andrade-Sanchez *et al.* (2013) and other crops (Furbank and Tester, 2011). Ghanem *et al.* (2015) also suggested that these approaches contain methods that obtain evidence that a hypothesized trait will lead to improvement and that phenotypic screens are multi-tiered so that insights about trait expression are gained at various stages of the breeding process.

5.2. Soil Management

Microbial processes have a central role in nutrient cycling, and hence are the key determinant of nutrient availability and nitrogen use efficiency in arable fields. Microbes are likely to respond rapidly to climate change. Whether changes in microbial processes lead to a net positive or negative impact on nutrient cycling remains debatable and will depend on soil nutritional status, soil types and climate conditions. However, understanding mechanisms and magnitude of nutrient cycling response to climate change is important in order to develop an effective adaptations strategy to minimize impacts on farm productivity and profitability. This involves consideration of the complex interactions that occur between microorganisms and other biotic and abiotic factors. The potential to manage and even enhance nutrient cycling under future climate through managing terrestrial microbial processes is a tantalizing prospect for the future, but significant research is needed to achieve this goal (Singh *et al.*, 2010), including:

- Approaches to building resilience in fields by increasing soil organic matter: a previous section of this review has addressed issues such as use of residue management, cover crops and minimum tillage to improve soil organic matter, which not only is a key nutritional source for both plants and microbes but also improves WUE by holding water for longer periods in root zones. Future research is needed to determine whether benefits of these approaches will be maintained under future climatic conditions. Research and innovations are also needed to identify other approaches, which can be harnessed to build soil organic matter in farmlands. This may include incorporation of bio-inoculants, external organic matters and biodegradable polymers that have potential to promote accumulation of soil organic matter in soils.
- Harnessing soil microflora to improve resource-use efficiency: soil microflora have a central role in determining resource availability and resource use efficiency. Previous studies have reported that biologically rich soils improves both resource and nutrient use efficiency (NUE) and hence farm productivity and profitability. Improved nitrogen use efficiency in biologically rich soils was achieved by continuous release and reduced loss of N and P in the soil profile (Bender and van der Heijden, 2015). Further research is needed to identify key microbial populations, which contribute to NUE, and factors, which determine their activities under current and future climatic conditions.
- Responses of key functional populations: functional groups which are directly linked to nitrogen (N-mineralizing, N-fixing communities, nitrifiers, denitrifiers) and phosphorus (P-mineralizing and P-solubilizing communities) availability and their response to climate change including extreme weathers is not fully understood. Both magnitude and direction of their response will be critical to develop nutrient management strategies, as their abundance, diversity and activity will ultimately determine the fate of applied as well as naturally available N and P in the farmland.
- Utility of bio-inoculants and biostimulants: it is proposed that addition of bio-inoculants and biostimulants could increase activities of beneficial microbes. One such specific approach includes utilization of mixed microbial inoculants (e.g. plant growth promoting bacteria and mycorrhiza), which is proposed to have better agronomic outcomes because this allows multiple mechanisms of resource-use efficiency to work simultaneously (Dodd and Ruiz-Lozano, 2012). However, our current understanding of survival of bio-inoculants in field conditions

and their interactions with crops is limited. Future research is also needed to identify how these bio-inoculants respond to future climatic conditions and how their interactions with the crop will be modified by climate change. Outcomes of such study will ultimately determine if such an approach can be effectively incorporated into climate adaptation strategies.

- Exploring rhizosphere–microbial interactions: harnessing rhizosphere–microbial interactions holds great promise to increase farm productivity under current and future climatic conditions. Previous studies have documented potential approaches to harness microbe–rhizosphere interactions for increased farm productivity (Shen *et al.*, 2013; Macdonald and Singh, 2014) but how these interactions could be exploited for biotechnological applications is not well defined. Plant roots and microbes communicate via chemical molecules for their energy and nutrient requirements and modulate each other's activities to achieve mutually beneficial outcomes. A major research effort is needed for identifying these signal molecules and harnessing them to improve interaction between beneficial microbes and plant roots that lead to improved resource availability and use efficiency. Once identified, these molecules can either be directly used or transgenic technologies can be used to genetically modify plants to manipulate microbial activities in the root zone, which can serve this purpose.

5.3. Cotton System Management

Climate change is a multifaceted and complex challenge for cotton industries and it will affect the sustainability of farms, ecosystems and the wider community. Fortunately, many potential adaptation responses available have immediate production efficiency benefits making them attractive options regardless of the rate and nature of future climate change. The following sections review potential challenges that climate change presents to the viability of cotton industries and identify key research activities needed to address these challenges.

5.3.1. Climate information and use

Climate change is occurring against a background of naturally high climate variability. For each cotton region it will be important to distinguish between climate variability and climate change as there is the potential for maladaptation, if not identified clearly. Adaptation in farming practice may respond and change to short-term variations in climate, which is not indicative of overall climate change, thereby leaving an industry vulnerable in the longer term. It will be also important for the provision of information on the likely impacts at the business level (i.e. downscaling climate change predictions to regional scales) and provide tools and extension networks to enable farmers to access climate data, and subsequently interpret data in relation to their crop records and analyse alternative management options. Increased forecasting skill in weather for prediction of extreme events such as heatwaves impacting plant growth, and heavy and excessive rainfall which can impact plant growth and soil erosion, will be vital. Crop simulation models can play an important role in assessing the skill and value of weather forecasts if decisions are changed in response to the forecast (McIntosh *et al.*, 2015).

5.3.2. Policy and industry considerations

The impacts of climate change and the approaches to adaptation will need to reflect changing social, political and economic drivers at scales that move from the field, to the farm, across varied agriculture industries, and with national and international influences. As an example, there is a need to invest in field-based research into production, but concurrently research is needed that assists in government policy setting. Without these types of considerations, the marginal return on investment into adaptation options can be severely diminished. Key considerations that capture some of these issues from a cotton production perspective include the following:

- An assessment of the likely impacts of climate change on worldwide cotton production. Understanding these impacts is necessary for maintaining the cotton supply chain to maintain market share security against synthetic textile production. Strengthening information sharing networks on impacts and adaptations to change will be vital to assist this process.

- Identifying opportunities for the expansion of cotton production in existing and new agriculture production regions. Region-specific impacts will need to be assessed so that cotton growers can improve their capacity to assess likely impacts at their business level.
- Identifying competition and synergies for use of resources (e.g. land, water, labour) from other agriculture enterprises. There is a need to address the question of just how much climate change it would take to make it more appropriate to consider using land and water resources for purposes other than cotton or irrigated production.
- Integrating research outcomes that are optimal in delivering sustainable cotton systems in light of triple bottom line (environmental, economic, and social) concerns. An example here is the need to develop a practice that is accepted in minimizing environmental concerns whilst optimizing grower profit. Some recent factors influencing this are the need for reductions in water accessibility for irrigation to meet environmental river flows and the need for improved N management (with improved timing, rate and use of legumes) to minimize GHG emissions and water contamination. Implementation of how these are managed can be sometimes in the form of government regulation, or the adaption of industry environmental management systems. In the case of government regulation influencing grower practice, it does serve to highlight that an investment in a production practice tailored and demonstrated to improve productivity can be simply overridden by government regulation influenced by other societal, environmental and economic concerns outside cotton production.
- Development of multi-peril crop insurance schemes to assist cotton producers deal with extreme climate events.

To meet these challenges there will be a greater need to incorporate other concepts of production use efficiencies into the analysis of modern cotton systems; e.g. fuel or energy use or carbon emissions per unit of lint produced, in addition to existing production use efficiencies (e.g. water and N). There will need to be 'trade-offs' to minimize economic, social and environmental harm, while maximizing new opportunities. One example that highlights this tension is the ever increasing need for WUE, which has led to demand for more sophisticated irrigation systems that are ultimately more energy intensive. Importantly to assist in making valid and fair comparisons within the cotton production system and beyond (e.g. with other cropping systems or industries), it will be necessary to present these efficiencies on an economic basis (e.g. $ generated /Ml, unit of GHG emitted, kg of N applied etc.).

Ultimately, sustainable and low environmental impact cotton systems are required to maintain modern cotton production systems' 'licence to farm'. Research into the development of new technology and tools that integrate knowledge at many scales, whilst understanding the linkages of on-farm production with the off-farm impacts, will be needed to reliably harness opportunities for ongoing investment.

5.3.3. Crop management

Considerable changes in climate may necessitate a reassessment of cotton systems to ensure maximum sustainability and economic return with all available resources. Strategies to mitigate damage incurred when encountering episodes of extreme environmental stress (through tolerance or avoidance) will need to be developed, building adaptive capacity and resilience. While formulating these strategies, they must take into account all aspects of the production system from planting through to harvest, and consider all the possible tools available (precision technologies and new genetics for example).

To help build resilient and productive systems a knowledge of yield potential or 'yield gap' in different cotton systems across regions will be important. This will assist in identifying the major limiting factors in systems and will also gain insights into the approaches to overcome these limitations. Importantly, these limitations require reassessment in light of future climate change predictions so that changes to the systems are not short-lived or maladaptive in the future. In many cases, the reduction in the 'yield gap' between farm averages and yield potential will be achieved more likely by removing yield constraints of the poorest fields and systems (Constable and Bange, 2015; Hatfield and Walthall, 2015). Crop simulation tools will play a key role in assessment of yield gaps. For rainfed systems, Hochman *et al.* (2012) measured farmers' yield and compared that with the regional yields predicted using simulation models of an adapted crop without limitations, but under water-limited conditions. They also assessed yield potential using crop competition results. For irrigated crop comparisons, it will be important that particular knowledge of the amount of water available for irrigation across farms is considered because it can vary considerably, thereby strongly affecting yield and fibre quality.

However, one of the most significant challenges for cotton management into the future will be diminished access to water through reductions in sources of irrigation (surface or groundwater), less rainfall, or increases in evapotranspiration through temperature increases. [eCO_2] increased WUE (kg lint/mm evapotranspiration) by increasing biomass production rather than by a reduction in consumptive use (Mauney et al., 1994). Much existing research has been undertaken in unlimited water conditions, and limited research has considered the implications of cotton growth and yield in current and future conditions with the same amount of water availability. Integrated climate change studies for cotton grown under varied [CO_2] and temperature regimes for irrigated cotton are required to be undertaken with different amounts of water availability that result in season-long reductions in transpiration compared to fully irrigated crops. For rainfed cotton systems, they will require closer examination of the response to various water deficits and drought recovery cycles. These effects will also need to be considered in light of other management options suggested in this review that relate to water use: the development of cotton systems that are earlier maturing, that use less water and allow more crops to be grown in rotation; and improved management options in limited water situations utilizing changes in planting time, alternative irrigation systems, row configurations, irrigation scheduling strategies, all with the intent to maximize WUE and maintain fibre quality.

One of the key challenges in undertaking research will be the development of large scale CO_2 enrichment facilities. Ziska et al. (2012) summarized engineering challenges in meeting this need and the limitation of various facilities have been discussed in detail earlier in this review. Briefly, most enclosure systems (various sorts of outdoor growth chambers) may suffer from 'chamber effects' such as light quality and/or quality issues associated with various types of chamber wall materials (c.f. Kim et al., 2004). On the other hand, FACE methods are known to have CO_2 control problems with large pulses of injected CO_2 over short durations of time resulting in an artificial lowering of plant responses to the apparent target [CO_2] set point (Bunce, 2011, 2014). The solution to these types of engineering challenges may be a combination of methods including large open-top chamber systems, environmentally controlled glasshouses and/or the utilization of naturally occurring CO_2 springs found in some parts of the world (Miglietta et al., 1993).

Although the impacts of [eCO_2], warmer temperatures and water deficits on cotton growth and physiology have been studied, gaps remain in our understanding of whether there are interactive relationships between these variables. It is important to understand potential interactions, as it is likely that multiple variables will be altered with future climatic changes. As mentioned throughout this review, modelling will play a vital role in quantifying the potential integrated effects of future climate change on physiology and growth of cotton, and translating these impacts into cotton production systems. They will also be important in evaluating the effectiveness of adaptation and mitigation strategies in dealing with climate change risk or in seeking to take advantage of climate change.

Models can be used to predict leaf and plant responses to changing conditions. The Farquhar et al. (1980) model was based on the kinetics of Rubisco, and it has been widely used for predicting the response of photosynthetic CO_2 fixation, and thus plant biomass production, to environmental change (Crafts-Brandner and Salvucci, 2004). The activation state of Rubisco is the primary metabolic limitation to photosynthetic CO_2 fixation under conditions of high temperature and high atmospheric [CO_2] (Crafts-Brandner and Salvucci, 2004). It will be important that these responses are tested in field conditions and that frameworks are developed that scale leaf and plant responses to canopy level responses (e.g. Hammer and Wright, 1994) to enable yield predictions.

There are a number of cotton simulation models that simulate climate change effects including [eCO_2] and published examples where they have been applied to investigate impacts and management (Thorp et al., 2014). For example, Reddy et al. (2002) quantified the effects of climate change on cotton production in the Mississippi Delta (USA) by using the cotton simulation model GOSSYM (Baker et al., 1983) with [eCO_2]. More recent studies by Luo et al. (2014, 2015) quantified the impact of climate change on cotton phenology, fibre quality, water use and yield using various empirical relationships as well as the use of OZCOT cotton simulation model (Hearn, 1994). In these studies, models were coupled with dynamically downscaled regional outputs of four global climate models. Modifications of both these models have principally focused on the modifications of canopy photosynthesis response to [eCO_2] using the same response (Reddy et al., 2008).

While the outcomes of these simulation studies make sense with our current understanding of cotton growth and physiology, it is however widely recognized that all crop simulation models still require considerable validation and quantification of the integrated climate change responses for field conditions, where yield and fibre quality are predicted, which is not a simple task. This is further complicated by the need to include changes in management and cultivars. In a recent global review of cotton simulation models for cotton systems, Thorp et al. (2014) suggest two needs for cotton simulation modelling to progress. The first is to compare existing cotton simulation models to identify their strengths and weaknesses. This could be achieved through modelling efforts being engaged in the Agricultural Model Intercomparison and

Improvement Project (AgMIP) (Thorp *et al.*, 2014; Schiermeier, 2015). The second need is to form multidisciplinary teams in the areas of climate science, crop science, computer science, and economics to improve, validate and apply these models. These approaches could specifically be applied to address both impacts and adaptation options for climate change for many of the concerns listed in this review.

VI CONCLUSION

Future climate change will impact cotton production systems; however, there will be opportunities to adapt. This review begins to provide details for the formation of robust frameworks to evaluate the impact of projected climatic changes, highlight the risks and opportunities with adaptation, and detail the approaches for investment in research. Major matters that were identified were:

- Climate change will have both positive and negative effects on cotton. Increased [CO_2] may increase yield in well-watered crops, and higher temperatures will extend the length of growing season (especially in current short-season areas). However, higher temperatures also have the potential to cause significant fruit loss, lower yields and alter fibre quality, and reduced water use efficiencies. Extreme weather events such as droughts, heatwaves and flooding also pose significant risks to improvements in cotton productivity.
- Research into integrated effects of climate change (temperature, humidity, [CO_2] and water stress) on cotton growth, yield and quality will require further investment. This includes the development of cultivars tolerant to abiotic stress (especially for more frequent hot, water-deficit and waterlogged situations). Some consideration or allowance will be needed in these studies for both cotton cultivars and insect pests that have been naturally selected in rising CO_2 environments.
- Although cotton is already well adapted to hot climates, continued breeding by conventional means as well as applying biotechnology tools and traits will develop cultivars with improved water use efficiency and heat tolerance. Investment in whole-plant and crop physiology will be important to develop robust understanding of the physiological determinants of cotton crop growth and development. Undertaking this research with the involvement of agronomic researchers, extension specialists, crop managers and growers is vital so that achievements can be recognized in the field as quickly as possible.
- The potential for declining availability of water resources under climate change will increase competition for these resources between irrigated cotton production, other crops and environmental uses. These issues emphasize the need for continual improvement in whole farm and crop water use efficiencies and the need for clear information on water availability.
- There will be a need to improve cotton farm resilience by maintaining and increasing cotton profitability through practices that increase both yield and fibre quality, while improving efficiency of resource use (especially energy, water and nitrogen).
- Region-specific effects will need to be assessed thoroughly so that cotton growers can assess likely impacts at the business level. Also, as cotton is a global commodity it will be vital for cotton industries to understand global changes in cotton markets as part of their overall adaptation strategy.
- Simulation models will play a vital role in assessing impacts and adaptation options for future climate change; however, they will require investment in development and their validation for climate change issues. As new forecasted future climate change scenarios are developed, there will also be a need to update and quantify impacts and re-evaluate adaptation options. Crop biophysical modelling should be appropriately linked to economic whole farm/catchment scale modelling efforts. Similar considerations need to be given to cotton decision support tools that utilize day-degree functions. It is possible that many systems do not accommodate future predicted extremes associated with climate change (e.g. heatwaves slowing crop development).
- Implementation of whole-farm designs that build system resilience through diversity in crops, while increasing soil fertility and protection from erosion through the inclusion of rotation and cover crops will also need further attention.

We acknowledge that most approaches discussed here are decidedly production focused, and therefore the list is by no means comprehensive. There are other significant efforts to combat 'a changing climate' from other perspectives and scales, policy and catchment scale efforts being examples. Ultimately, it is a multi-faceted systems-based approach that combines all elements mentioned here as well as others that provide the best insurance to harness the change that is occurring, and best allow cotton industries to adapt. Given that there will be no single solution for all of the challenges raised by climate change and variability, the best adaptation strategy for industry will be to develop more resilient systems. Early implementation of adaptation strategies, particularly in regard to enhancing resilience, has the potential to significantly reduce the negative impacts of climate change.

Abbot, A.M., Hequet, E.F., Higgerson, G.J., Lucas, S.R., Naylor, G.R.S., Purmalis, M.M. and Thibodeaux, D.P. (2011) Performance of the Cottonscan (TM) instrument for measuring the average fiber linear density (fineness) of cotton lint samples. *Textile Research Journal* 81, 94–100.

Acosta-Martinez, V., Cotton, J., Gardner, T., Moore-Kucera, J., Zak, J., Wester, D. and Cox, S. (2014a) Predominant bacterial and fungal assemblages in agricultural soils during a record drought/heat wave and linkages to enzyme activities of biogeochemical cycling. *Applied Soil Ecology* 84, 69–82.

Acosta-Martinez, V., Moore-Kucera, J., Cotton, J., Gardner, T. and Wester, D. (2014b) Soil enzyme activities during the 2011 Texas record drought/heat wave and implications to biogeochemical cycling and organic matter dynamics. *Applied Soil Ecology* 75, 43–51.

Aerts, R. (2006) The freezer defrosting: global warming and litter decomposition rates in cold biomes. *Journal of Ecology* 94, 713–724.

Ainsworth, E.A. and Long, S.P. (2005) What have we learned from 15 years of free-air CO_2 enrichment (FACE)? A meta-analytic review of the responses of photosynthesis, canopy properties and plant production to rising CO_2. *New Phytologist* 165, 351–371.

Ainsworth, E.A. and McGrath, J.M. (2010) Climate Change and Food Security Direct Effects of Rising Atmospheric Carbon Dioxide and Ozone on Crop Yields. In: Lobell, D. and Burke M. (eds) *Climate Change and Food Security: Adapting Agriculture to a Warmer World*. Springer, New York, pp. 109–130.

Ainsworth, E.A. and Rogers, A. (2007) The response of photosynthesis and stomatal conductance to rising CO_2: mechanisms and environmental interactions. *Plant Cell and Environment* 30, 258–270.

Allen, R.D. and Aleman, L. (2011) Abiotic stress and cotton fiber development. In: Oosterhuis, D.M. (ed.) *Stress Physiology in Cotton*. The Cotton Foundation, Cordova, Tennessee.

Amthor, J.S. (1991) Respiration in a future, higher-CO2 world. *Plant, Cell & Environment* 14, 13–20.

Amthor, J.S. (1997) Plant respiratory responses to elevated carbon dioxide partial pressure. In: Allen, L.H. Jr. *et al.* (eds) Advances in carbon dioxide effects research. *Proceedings of a symposium*, Cincinnati, Ohio, USA, 7–12 November 1993, pp. 35–77.

Amthor, J.S. (2000) Direct effect of elevated CO_2 on nocturnal in situ leaf respiration in nine temperate deciduous tree species is small. *Tree Physiology* 20, 139–144.

Anderson, T.H., Heinemeyer, O. and Weigel, H.J. (2011) Changes in the fungal-to-bacterial respiratory ratio and microbial biomass in agriculturally managed soils under free-air CO_2 enrichment (FACE) – A six-year survey of a field study. *Soil Biology & Biochemistry* 43, 895–904.

Andrade-Sanchez, P., Gore, M.A., Heun, J.T., Thorp, K.R., Carmo-Silva, A.E., French, A.N., Salvucci, M.E. and White, J.W. (2013) Development and evaluation of a field-based high-throughput phenotyping platform. *Functional Plant Biology* 41, 68–79.

Arevalo, M., Oosterhuis, D.M., Coker, D. and Brown, R. (2008) Physiological response of cotton to high night temperature. *American Journal of Plant Science and Biotechnology* 2, 63–68.

Arndt, C.H. (1945) Temperature-growth relations of the roots and hypocotyls of cotton seedlings. *Plant Physiology* 20, 200–220.

Arp, W.J. (1991) Effects of source-sink relations on photosynthetic acclimation to elevated CO_2. *Plant, Cell and Environment* 14, 869–875.

Arriaga, F.J., Prior, S.A., Terra, J.F. and Delaney, D.P. (2009) Cotton gas exchange response to standard and ultra-narrow row systems under conventional and no-tillage. *Communications in Biometry and Crop Science* 4, 42–51.

Assmann, S.M. (1993) Signal Transduction in Guard Cells. *Annual Review of Cell Biology* 9, 345–375.

Assmann, S.M. (1999) The cellular basis of guard cell sensing of rising CO_2. *Plant Cell and Environment* 22, 629–637.

Baggs, E.M., Richter, M., Cadisch, G. and Hartwig, U.A. (2003) Denitrification in grass swards is increased under elevated atmospheric CO_2. *Soil Biology & Biochemistry* 35, 729–732.

Bai, E., Li, S.L., Xu, W.H., Li, W., Dai, W.W. and Jiang, P. (2013) A meta-analysis of experimental warming effects on terrestrial nitrogen pools and dynamics. *New Phytologist* 199, 441–451.

Baker, D.N. (1965) Effects of certain environmental factors on net assimilation in cotton. *Crop Science* 5, 53–56.

Baker, D.N. and Hesketh, J.D. (1969) Respiration and the carbon balance in cotton (*Gossypium hirsutum* L.). *Beltwide Cotton Production Research Conferences*, National Cotton Council, Memphis, New Orleans, pp. 60–64.

Baker, D.N., Hesketh, J.D. and Duncan, W.G. (1972) Simulation of growth and yield in cotton. 1. Gross photosynthesis, respiration, and growth. *Crop Science* 12, 431–435.

Baker, D.N., Lambert, J.R. and McKinion, J.M. (1983) GOSSYM: A simulator of cotton crop growth and yield. *South Carolina Agricultural Experiment Station Technical Bulletin* 1086, Clemson University, Clemson, South Carolina.

Baker, J.T. (2004) Yield responses of southern US rice cultivars to CO_2 and temperature. *Agricultural and Forest Meteorology* 122, 129–137.

Baker, J.T. and Allen, L.H. (2005) Rice Growth, Yield and Photosynthetic Responses to Elevated Atmospheric Carbon Dioxide Concentration and Drought. *Journal of Crop Improvement* 13, 7–30.

Baker, J.T., Allen, L.H., Boote, K.J., Jones, P. and Jones, J.W. (1989) Response of soybean to air-temperature and carbon-dioxide concentration. *Crop Science* 29, 98–105.

Baker, J.T., Allen, L.H., Boote, K.J., Jones, P. and Jones, J.W. (1990) Rice ,
ation in subambient, ambient, and superambient carbon-dioxide concentrations.
834–840.

Baker, J.T., Gitz, D.C., Payton, P., Wanjura, D.F. and Upchurch, D.R. (2007) Using leaf gas exchange to quantify drought in cotton irrigated based on canopy temperature measurements. *Agronomy Journal* 99, 637–644.

Baker, J.T., Van Pelt, S., Gitz, D.C., Payton, P., Lascano, R.J. and McMichael, B. (2009) Canopy Gas Exchange Measurements of Cotton in an Open System. *Agronomy Journal* 101, 52–59.

Baker, J.T., Gitz, D.C. and Lascano, R.J. (2014a) Field Evaluation of Open System Chambers for Measuring Whole Canopy Gas Exchanges. *Agronomy Journal* 106, 537–544.

Baker, J.T., Gitz, D.C., Payton, P., Broughton, K.J., Bange, M.P. and Lascano, R.J. (2014b) Carbon Dioxide Control in an Open System that Measures Canopy Gas Exchanges. *Agronomy Journal* 106, 789–792.

Bange, M.P. and Milroy, S.P. (2001) Effect of temperature on the rate of early fruiting developmental processes of cotton. *10th Australian Agronomy Conference*, Australian Society of Agronomy, Hobart, Tasmania.

Bange, M.P. and Milroy, S.P. (2004) Growth and dry matter partitioning of diverse cotton genotypes. *Field Crops Research* 87, 73–87.

Bange, M., Carberry, P. and Hammer, G. (1999) Seasonal climate forecasting for cotton management. *The Australian Cottongrower* 20, 36–42.

Bange, M.P., Milroy, S.P. and Thongbai, P. (2004) Growth and yield of cotton in response to waterlogging. *Field Crops Research* 88, 129–142.

Bange, M.P., Carberry, P.S., Marshall, J. and Milroy, S.P. (2005) Row configuration as a tool for managing rain-fed cotton systems: review and simulation analysis. *Australian Journal of Experimental Agriculture* 45, 65–77.

Bange, M., Milroy, S. and Roberts, G. (2006) Factors influencing crop maturity. *Beltwide Cotton Conference*, San Antonio, Texas.

Bange, M.P., Caton, S.J. and Milroy, S.P. (2008) Managing yields of high fruit retention in transgenic cotton (*Gossypium hirsutum* L.) using sowing date. *Australian Journal of Agricultural Research* 59, 733–741.

Bange, M.P., Constable, G.A., Johnston, D.B. and Kelly, D. (2010a) A method to estimate the effects of temperature on cotton micronaire. *Journal of Cotton Science* 14, 164–172.

Bange, M.P., Constable, G.A., McRae, D. and Roth, G. (2010b) Cotton. In: Stokes, C. and Howden, M. (eds) *Adapting Agriculture to Climate Change: Preparing Australian Agriculture, Forestry and Fisheries for the Future*. CSIRO Publishing, Melbourne, Australia, pp. 49–66.

Bange, M.P., Milroy, S.P., Ellis, M.H. and Thongbai, P. (2010c) Opportunities to reduce the impact of water-logging on cotton. *15th Australian Agronomy Conference*, Australian Society of Agronomy, Christchurch, New Zealand.

Bapiri, A., Baath, E. and Rousk, J. (2010) Drying-rewetting cycles affect fungal and bacterial growth differently in an arable soil. *Microbial Ecology* 60, 419–428.

Barnard, R., Leadley, P.W. and Hungate, B.A. (2005) Global change, nitrification, and denitrification: a review. *Global Biogeochemical Cycles* 19.

Barrow, J.R. (1983) Comparisons among pollen viability measurement methods in cotton. *Crop Science* 23, 734–736.

Bauer, P.J. and Busscher, W.J. (1996) Winter cover and tillage influences on coastal plain cotton production. *Journal of Production Agriculture* 9, 50–54.

Bednarz, C.W. and van Iersel, M.W. (2001) Temperature response of whole-plant CO_2 exchange rates of four upland cotton cultivars differing in leaf shape and leaf pubescence. *Communications in Soil Science and Plant Analysis* 32, 2485–2501.

Bell, C.W., Acosta-Martinez, V., McIntyre, N.E., Cox, S., Tissue, D.T. and Zak, J.C. (2009) Linking microbial community structure and function to seasonal differences in soil moisture and temperature in a Chihua-huan Desert grassland. *Microbial Ecology* 58, 827–842.

Bender, S.F. and van der Heijden, M.G.A. (2015) Soil biota enhance agricultural sustainability by improving crop yield, nutrient uptake and reducing nitrogen leaching losses. *Journal of Applied Ecology* 52, 228–239.

Berard, A., Bouchet, T., Sevenier, G., Pablo, A.L. and Gros, R. (2011) Resilience of soil microbial commu-nities impacted by severe drought and high temperature in the context of Mediterranean heat waves. *European Journal of Soil Biology* 47, 333–342.

Berry, J. and Bjorkman, O. (1980) Photosynthetic response and adaptation to temperature in higher-plants. *Annual Review of Plant Physiology and Plant Molecular Biology* 31, 491–543.

Besford, R.T., Ludwig, L.J. and Withers, A.C. (1990) The greenhouse-effect – acclimation of tomato plants growing in high CO2, photosynthesis and ribulose-1,5-bisphosphate carboxylase protein. *Journal of Experimental Botany* 41, 925–931.

Bhattacharya, N.C., Radin, J.W., Kimball, B.A., Mauney, J.R., Hendrey, G.R., Nagy, J., Lewin, K.F. and Ponce, D.C. (1994) Leaf water relations of cotton in a free-air CO_2-enriched environment. *Agricultural and Forest Meteorology* 70, 171–182.

Bibi, A.C., Oosterhuis, D.M. and Gonias, E.G. (2008a) Photosynthesis, quantum yield of photosystem II, and membrane leakage as affected by high temperatures in cotton genotypes. *Jounal of Cotton Science* 12, 150–159.

Bibi, C., Oosterhuis, D.M. and Gonias, E.D. (2008b) Changes in the antioxidant enzymes activity of cotton genotypes during high temperature stress. *Life Sciences International Journal* 2, 621–627.

Bibi, A.C., Oosterhuis, D.M. and Gonias, E.D. (2010a) Exogenous application of putrescine ameliorates the effect of high temperature in *Gossypium hirsutum* L. flowers and fruit development. *Journal of Agronomy and Crop Science* 196, 205–211.

Bibi, A.C., Oosterhuis, D.M., Gonias, E.D. and Stewart, J.M. (2010b) Comparison of a responses of a ruderal *Gossypium hirsutum* L. with commercial cotton genotypes under high temperature stress. *American Journal of Plant Science and Biotechnology* 4, 87–92.

Blankenship, S.M.D. and Dole, J.M. (2003) 1-Methylcyclopropene: a review. *Postharvest Biology and Technology* 28, 1–25.

Blankinship, J.C., Niklaus, P.A. and Hungate, B.A. (2010) A meta-analysis of responses of soil biota to global change. *Oecologia* 165, 553–565.

Bordovsky, J.P., Lyle, W.M., Lascano, R.J. and Upchurch, D.R. (1992) Cotton irrigation management with LEPA systems. *Transactions of the ASAE* 44, 309–332.

Bordovsky, J.P., Lyle, W.M. and Keeling, J.W. (1994) Crop-rotation and tillage effects on soil-water and cotton yield. *Agronomy Journal* 86, 1–6.

Bowes, G. (1993) Facing the inevitable: Plants and increasing atmospheric carbon dioxide. In: Briggs, W.R. (ed.) *Annual Review of Plant Physiology and Plant Molecular Biology*. Annual Reviews, Palo Alto, California, pp. 309–332.

Bradow, J.M. and Davidonis, G.H. (2000) Quantitation of fiber quality and cotton production-processing interface: A physiologist's perspective. *The Journal of Cotton Science* 4, 34–64.

Bradow, J.M. and Davidonis, G.H. (2010) Effects of environment on fiber quality. In: Stewart, A.M., Oosterhuis, D.M., Heitholt, J.J. and Mauney, J.R. (eds) *Physiology of Cotton*. Springer, New York.

Braganza, K. and Church, J. (2011) Observations of Global and Australian Climate. In: Cleugh, H., Stafford Smith, M., Battaglia, M. and Graham, P. (eds) *Climate Change: Science and Solutions for Australia*. CSIRO, Victoria, Australia.

Braunack, M.V. and Johnston, D.B. (2014) Changes in soil cone resistance due to cotton picker traffic during harvest on Australian cotton soils. *Soil and Tillage Research* 140, 29–39.

Braunack, M.V., Bange, M.P. and Johnston, D.B. (2012) Can planting date and cultivar selection improve resource use efficiency of cotton systems? *Field Crops Research* 137, 1–11.

Braunack, M.V., Johnston, D.B., Price, J. and Gauthier, E. (2015) Soil temperature and soil water potential under thin oxodegradable plastic film impact on cotton crop establishment and yield. *Field Crops Research* 184, 91–103.

Brearley, J., Venis, M.A. and Blatt, M.R. (1997) The effect of elevated CO_2 concentrations on K^+ and anion channels of *Vicia faba* L. guard cells. *Planta* 203, 145–154.

Brito, G.G.d., Ferreira, A.C.d.B., Borin, A.L.D.C. and Morello, C.d.L. (2013) 1-methylcyclopropene and aminoethoxyvinylglycine effects on yield components of field-grown cotton. *Ciencia e Agrotecnologia* 37, 9–16.

Brodrick, R., Neilsen, J., Bange, M., Hodgson, D. and Mundey, L. (2012) Dynamic deficits for irrigated cotton – matching the soil water to plant requirements. *16th Australian Agronomy Conference*, Australian Society of Agronomy, Armidale, New South Wales.

Brook, K.D., Hearn, A.B. and Kelly, C.F. (1992) Response of cotton to damage by insects pests in Australia: Compensation for early season fruit damage. *Journal of Economic Entomology* 85, 1378–1386.

Broughton, K.J. (2015) The integrated effects of projected climate change on cotton growth and physiology. Thesis, Faculty of Agriculture and Environment, The University of Sydney.

Brown, S. and Oosterhuis, D.M. (2010) High daytime temperature stress effects on the physiology of modern versus obsolete cultivars. *American Journal of Plant Science and Biotechnology* 4, 93–96.

Bruce, R.R., Langdale, G.W., West, L.T. and Miller, W.P. (1995) Surface soil degradation and soil productivity restoration and maintenance. *Soil Science Society of America Journal* 59, 654–660.

Bu, L.-D., Liu, J.-L., Zhu, L., Luo, S.-S., Chen, X.-P., Li, S.-Q., Lee Hill, R. and Zhao, Y. (2013) The effects of mulching on maize growth, yield and water use in a semi-arid region. *Agricultural Water Management* 123, 71–78.

Bunce, J.A. (1983) Differential sensitivity to humidity of daily photosynthesis in the field in C3-species and C4-species. *Oecologia* 57, 262–265.

Bunce, J.A. (2011) Performance characteristics of an area distributed free air carbon dioxide enrichment (FACE) system. *Agricultural and Forest Meteorology* 151, 1152–1157.

Bunce, J.A. (2012) Responses of cotton and wheat photosynthesis and growth to cyclic variation in carbon dioxide concentration. *Photosynthetica* 50, 395–400.

Bunce, J.A. (2014) CO_2 enrichment at night affects the growth and yield of common beans. *Crop Science* 54, 1744–1747.

Bunce, J.A. and Nasyrov, M. (2012) A new method of applying a controlled soil water stress, and its effect on the growth of cotton and soybean seedlings at ambient and elevated carbon dioxide. *Environmental and Experimental Botany* 77, 165–169.

Bureau of Meteorology (2011) Australia Water Definition Dictionary: water pressure deficit. Available at: http://www.bom.gov.au/water/awid/id-569.shtml (accessed 24 November 2015).

Burke, J.J. and Upchurch, D.R. (1989) Leaf temperature and transpirational control in cotton. *Environmental and Experimental Botany* 29, 487–492.

Burke, J.J. and Wanjura, D.F. (2009) Plant responses to temperature extremes. In: Stewart, J.M., Oosterhuis, D.M., Heitholt, J.J. and Mauney, J.R. (eds) *Physiology of Cotton*. Springer, New York, pp. 123–128.

Burke, J.J., Mahan, J.R. and Hatfield, J.L. (1988) Crop-specific thermal kinetic windows in relation to wheat and cotton biomass production. *Agronomy Journal* 80, 553–556.

Burke, J.J., Velten, J. and Oliver, M.J. (2004) *In vitro* analysis of cotton pollen germination. *Agronomy Journal* 96, 359–368.

Campbell, B.T., Boykin, D., Abdo, Z. and Meredith, W.R. (2014) Cotton. In: Smith, S., Specht, J., Diers, B. and Carver, B. (eds) *Yield Gains in Major US Field Crops*. CSSA Special Publication ASA, CSSA, and SSSA, Madison, Wisconsin, pp. 13–32.

Carberry, P., Hammer, G., Meinke, H. and Bange, M. (2000) The potential value of seasonal climate forecasting in managing cropping systems. In: Hammer, G.L., Nicholls, N. and Mitchell, C. (eds) *Applications of Seasonal Climate Forecasting in Agricultural and Natural Ecosystems – the Australian Experience.* Springer, Dordrecht, the Netherlands, pp. 167–181.

Carberry, P.S., Bruce, S.E., Walcott, J.J. and Keating, B.A. (2011) Innovation and productivity in dryland agriculture: a return-risk analysis for Australia. *Journal of Agricultural Science* 149, 77–89.

Carnol, M., Hogenboom, L., Jach, M.E., Remacle, J. and Ceulemans, R. (2002) Elevated atmospheric CO_2 in open top chambers increases net nitrification and potential denitrification. *Global Change Biology* 8, 590–598.

Castro, H.F., Classen, A.T., Austin, E.E., Norby, R.J. and Schadt, C.W. (2010) Soil microbial community responses to multiple experimental climate change drivers. *Applied and Environmental Microbiology* 76, 999–1007.

Chakravarthy, V.S., Reddy, T.P., Reddy, V.D. and Rao, K.V. (2014) Current status of genetic engineering in cotton (*Gossypium hirsutum* L): an assessment. *Critical Reviews in Biotechnology* 34, 144–160.

Chaves, A.L.S. and Mello-Farias, P.C.d. (2006) Ethylene and fruit ripening: from illumination gas to the control of gene expression, more than a century of discoveries. *Genetics and Molecular Biology* 29, 508–515.

Chen, F.J., Wu, G., Parajulee, M.N. and Ge, F. (2007) Long-term impacts of elevated carbon dioxide and transgenic Bt cotton on performance and feeding of three generations of cotton bollworm. *Entomologia Experimentalis et Applicata* 124, 27–35.

China (2004) *The Peoples' Republic of China's Initial National Communication under the United Nations Framework Convention on Climate Change.* UNFCCC, Switzerland, 156 pp.

Christianson, J.A., Llewellyn, D.J., Dennis, E.S. and Wilson, I.W. (2010) Global gene expression responses to waterlogging in roots and leaves of cotton (*Gossypium hirsutum* L.). *Plant and Cell Physiology* 51, 21–37.

Clement, J.D., Constable, G.A., Stiller, W.N. and Liu, S.M. (2012) Negative associations still exist between yield and fibre quality in cotton breeding programs in Australia and USA. *Field Crops Research* 128, 1–7.

Clement, J.D., Constable, G.A., Stiller, W.N. and Liu, S.M. (2015) Early generation selection strategies for breeding better combinations of cotton yield and fibre quality. *Field Crops Research* 172, 145–152.

Conaty, W.C., Tan, D.K.Y., Constable, G.A., Sutton, B.G., Field, D.J. and Mamum, E.A. (2008) Genetic variation for waterlogging tolerance in cotton. *Journal of Cotton Science* 12, 53–61.

Conaty, W.C., Burke, J.J., Mahan, J.R., Neilsen, J.E. and Sutton, B.G. (2012) Determining the optimum plant temperature of cotton physiology and yield to improve plant-based irrigation scheduling. *Crop Science* 52, 1828–1836.

Conaty, W.C., Mahan, J.R., Neilsen, J.E. and Constable, G.A. (2014) Vapour pressure deficit aids the interpretation of cotton canopy temperature response to water deficit. *Functional Plant Biology* 41, 535–546.

Constable, G.A. (1991) Mapping the production and survival of fruit on field grown cotton. *Agronomy Journal* 83, 374–378.

Constable, G.A. (1995) Predicting yield responses of cotton to growth regulators. In: Constable, G.A. and Forrester, N.W. (eds) *World Cotton Research Conference – 1: Challenging the future.* CSIRO, Melbourne, Brisbane, Australia, pp. 6–24.

Constable, G.A. and Bange, M.P. (2008) Producing and preserving fiber quality: from the seed to the bale. In: *Proceedings of the 4th World Cotton Conference*, 10–14 September 2007, Lubbock, USA. International Cotton Advisory Committee, Washington, DC.

Constable, G.A. and Bange, M.P. (2015) The yield potential of cotton (*Gossypium hirsutum* L). *Field Crop Research* 182, 98–106.

Constable, G.A. and Hearn, A.B. (1981) Irrigation for crops in a sub-humid environment VI. Effect of irrigation and nitrogen fertliser on growth, yield and quality of cotton. *Irrigation Science* 3, 17–28.

Constable, G.A. and Rawson, H.M. (1980a) Carbon production and utilization in cotton – inferences from a carbon budget. *Australian Journal of Plant Physiology* 7, 539–553.

Constable, G.A. and Rawson, H.M. (1980b) Effect of leaf position, expansion and age on photosynthesis, transpiration and water use efficiency of cotton. *Australian Journal of Plant Physiology* 7, 89–100.

Constable, G.A., Harris, N.V. and Paull, R.E. (1976) The effect of planting date on the yield and some fibre properties of cotton in the Namoi Valley. *Australian Journal of Experimental Agriculture and Animal Husbandry* 16, 265–271.

Constable, G.A., Reid, P.E. and Thomson, N.J. (2001) *Approaches Utilized in Breeding and Development of Cotton Cultivars in Australia.* Science Publishers, Enfield, New Hampshire.

Cothren, J.T. (1995) Use of growth regulators in cotton production. In: Constable, G.A. and Forrester, N.W. (eds) *World Cotton Research Conference – 1: Challenging the future.* CSIRO, Melbourne, Brisbane, Australia, pp. 1–3.

Cottee, N.S., Tan, D.K.Y., Bange, M.P., Cothren, J.T. and Campbell, L.C. (2010) Multi-level determination of heat tolerance in cotton (*Gossypium hirsutum* L.) under field conditions. *Crop Science* 50, 2553–2564.

Coviella, C.E., Morgan, D.J.W. and Trumble, J.T. (2000) Interactions of elevated CO_2 and nitrogen fertilization: effects on production of *Bacillus thuringiensis* toxins in transgenic plants. *Environmental Entomology* 29, 781–787.

Coviella, C.E., Stipanovic, R.D. and Trumble, J.T. (2002) Plant allocation to defensive compounds: interactions between elevated CO_2 and nitrogen in transgenic cotton plants. *Journal of Experimental Botany* 53, 323–331.

Crafts-Brandner, S.J. and Salvucci, M.E. (2004) Analyzing the impact of high temperature and CO_2 on net photosynthesis: biochemical mechanisms, models and genomics. *CSSA Symposium 'Opportunities linking functional genomics with physiology for global change research'*, Denver, Colorado, 5 November 2003, pp. 75–85.

CSIRO, Bureau of Meteorology (2012) State of the Climate. Available at: http://www.csiro.au/.../LM/FOI%20Disclosure%20Log%202012-13/Climate%20Snapshot%202012%20Brochure.pdf?la=en (accessed 24 November 2015).

Cure, J.D. and Acock, B. (1986) Crop responces to carbon dioxide doubling – a literature survey. *Agricultural and Forest Meteorology* 38, 127–145.

Dai, J. and Dong, H. (2014) Intensive cotton farming technologies in China: Achievements, challenges and countermeasures. *Field Crops Research* 155, 99–110.

Davey, P.A., Parsons, A.J., Atkinson, L., Wadge, K. and Long, S.P. (1999) Does photosynthetic acclimation to elevated CO_2 increase photosynthetic nitrogen-use efficiency? A study of three native UK grassland species in open-top chambers. *Functional Ecology* 13, 21–28.

Davidson, E.A. and Janssens, I.A. (2006) Temperature sensitivity of soil carbon decomposition and feedbacks to climate change. *Nature* 440, 165–173.

Delucia, E., Sesek, T.W. and Strain, B.R. (1985) Photosynthetic inhibition after long-term exposure to elevated levels of atmospheric carbon dioxide. *Photosynthesis Research* 7, 175–184.

Diaz, S., Grime, J.P., Harris, J. and McPherson, E. (1993) Evidence of a feedback mechanism limiting plant-response to elevated carbon-dioxide. *Nature* 364, 616–617.

Djanaguiraman, M., Sheeba, J.A., Devi, D.D., Bangarusamy, U. and Prasad, P.V. (2010) Nitrophenolates spray can alter boll abscission rate in cotton through enhanced peroxidase activity and increased ascorbate and phenolics levels. *Journal of Plant Physiology* 167, 1–9.

Dodd, I.C. and Ruiz-Lozano, J.M. (2012) Microbial enhancement of crop resource use efficiency. *Current Opinion in Biotechnology* 23, 236–242.

Doherty, R.M., Mearns, L.O., Reddy, K.R., Downton, M.W. and McDaniel, L. (2003) Spatial scale effects of climate scenarios on simulated cotton production in the southeastern USA. *Climatic Change* 60, 99–129.

Dong, H., Ii, W., Tang, W., Li, Z., Zhang, D. and Niu, Y. (2006) Yield, quality and leaf senescence of cotton grown at varying planting dates and plant densities in the Yellow River Valley of China. *Field Crops Research* 98, 106–115.

Downton, J. and Slatyer, R.O. (1972) Temperature dependence of photosynthesis in cotton. *Plant Physiology* 50, 518–522.

Drake, B.G., Gonzalez-Meler, M.A. and Long, S.P. (1997) More efficient plants: A consequence of rising atmospheric CO_2? *Annual Review of Plant Physiology and Plant Molecular Biology* 48, 609–639.

Drew, M.C. (1997) Oxygen deficiency and root metabolsm: injury and acclimation under hypoxia and anoxia. *Annual Review of Plant Physiology and Plant Molecular Biology* 48, 223–250.

Dugas, W.A., Heuer, M.L., Hunsaker, D., Kimball, B.A., Lewin, K.F., Nagy, J. and Johnson, M. (1994) Sap flow measurements of transpiration from cotton grown under ambient and enriched CO_2 concentrations. *Agricultural and Forest Meteorology* 70, 231–245.

Duursma, R.A., Payton, P., Bange, M.P., Broughton, K.J., Smith, R.A., Medlyn, B.E. and Tissue, D.T. (2013) Near-optimal response of instantaneous transpiration efficiency to vapour pressure deficit, temperature and [CO_2] in cotton (*Gossypium hirsutum* L.). *Agricultural and Forest Meteorology* 168, 168–176.

Eaton, F.M. and Ergle, D.R. (1953) Effects of shade and partial defoliation on carbohydrate levels and the growth, fruiting and fiber properties of cotton plants. *Plant Physiology* 29, 39–49.

El-Sharkawy, M.A. and Hesketh, J.D. (1964) Effects of temperature and water deficit on leaf photosynthetic rates of different species. *Crop Science* 4, 514–518.

Ellis, M.H., Millar, A.A., Llewellyn, D.J., Peacock, W.J. and Dennis, E.S. (2000) Transgenic cotton (*Gossypium hirsutum*) over-expressing alcohol dehydrogenase shows increased ethanol fermentation but no increase in tolerance to oxygen deficiency. *Functional Plant Biology* 27, 1041–1050.

Ennahli, S. and Earl, H.J. (2005) Physiological limitations to photosynthetic carbon assimilation in cotton under water stress. *Crop Science* 45, 2374–2382.

Ephrath, J.E., Timlin, D.J., Reddy, V. and Baker, J. (2011) Irrigation and elevated carbon dioxide effects on whole canopy photosynthesis and water use efficiency in cotton (*Gossypium hirsutum* L.). *Plant Biosystems* 145, 202–215.

Estiarte, M., Penuelas, J., Kimball, B.A., Idso, S.B., Lamorte, R.L., Pinter, P.J., Wall, G.W. and Garcia, R.L. (1994) Elevated CO_2 effects on stomatal density of wheat and sour orange trees. *Journal of Experimental Botany* 45, 1665–1668.

Evans, L.S. and Hendrey, G.R. (1992) Responses of cotton foliage to short-term fluctuations in CO_2 partial pressures. *Critical Reviews in Plant Sciences* 11, 203–212.

Evenson, J.P. (1969) Effects of floral and terminal bud removal on the yield and structure of the cotton plant in the Ord Valley, North Western Australia. *Cotton Growing Review* 46, 37–44.

Farquhar, G.D. and Sharkey, T.D. (1982) Stomatal conductance and photosynthesis. *Annual Review of Plant Physiology* 33, 317–345.

Farquhar, G.D., Caemmerer, S.v. and Berry, J.A. (1980) A biochemical model of photosynthetic CO_2 assimilation in leaves of C_3 species. *Planta* 149, 78–90.

Fitt, G.P. (2000) An Australian approach to IPM in cotton: integrating new technologies to minimise insecticide dependance. *Crop Protection* 19, 793–800.

Fitt, G.P. and Wilson, L.J. (2000) Genetic Engineering in IPM: Bt cotton. In: Kennedy, G.G. and Sutton, T.B. (eds) *Emerging Technologies in Integrated Pest Management: Concepts, Research and Implementation.* APS Press, St Paul, Minnesota, pp. 108–125.

Fitt, G.P., Dillion, M.L. and Hamilton, J.G. (1995) Spatial distribution of *Helicoverpa* populations in Australia: simulation modelling and empirical studies of adult movement. *Computers and Electronics in Agriculture* 13, 177–192.

49

Fleisher, D.H., Timlin, D.J., Yang, Y., Reddy, V.R. and Reddy, K.R. (2009) Uniformity of soil-plant-atmosphere-research chambers. *Transactions of the ASABE* 52, 1721–1731.

Fleisher, D.H., Timlin, D.J., Yang, Y., Reddy, V.R. and Reddy, K.R. (2010) Effects of carbon dioxide and temperature on crops: Lessons from SPAR growth chambers. In: Hillel, D. and Rosenzweig, K.F. (eds) *Handbook of Climate Change and Agroecosystems: Impacts, Adaptation, and Mitigation.* Imperial College Press, London, pp. 55–86.

Flexas, J. and Medrano, H. (2002) Drought-inhibition of photosynthesis in C_3 plants: Stomatal and non-stomatal limitations revisited. *Annals of Botany* 89, 183–189.

Fuchslueger, L., Bahn, M., Fritz, K., Hasibeder, R. and Richter, A. (2014) Experimental drought reduces the transfer of recently fixed plant carbon to soil microbes and alters the bacterial community composition in a mountain meadow. *New Phytologist* 201, 916–927.

Furbank, R.T. and Tester, M. (2011) Phenomics – technologies to relieve the phenotyping bottleneck. *Trends in Plant Science* 16, 635–644.

Ghanem, M.E., Marrou, H. and Sinclair, T.R. (2015) Physiological phenotyping of plants for crop improvement. *Trends in Plant Science* 20, 139–144.

Gipson, J.R. and Joham, H.E. (1968a) Influence of night temperature on growth and development of cotton (*Gossypium hirsutum* L.). II. Fiber properties. *Agronomy Journal* 60, 296–298.

Gipson, J.R. and Joham, H.E. (1968b) Influence of night temperature on growth and development of cotton (*Gossypium hirsutum* L.) I. Fruiting and boll development. *Agronomy Journal* 60, 292–295.

Gipson, J.R. and Ray, L.L. (1970) Temperature variety inter-relationships in cotton. I. Boll and fiber development. *Cotton Growing Review* 47, 257–271.

Goransson, H., Godbold, D.L., Jones, D.L. and Rousk, J. (2013) Bacterial growth and respiration responses upon rewetting dry forest soils: Impact of drought-legacy. *Soil Biology and Biochemistry* 57, 477–486.

Goswami, C.L. (1990) Hormonal regulation of fertility-induced changes in stomatal diffusive resistance in water-logged cotton (*Gossypium hirsutum* L.) var H-777. *Indian Journal of Experimental Biology* 28, 585–587.

Grace, P.R., Rowlings, D., Rochester, I.J., Kiese, R. and Butterbach-Bahl, K. (2010) Nitrous oxide emissions from irrigated cotton soils of northern Australia. *19th World Congress of Soil Science, Soil Solutions for a Changing World*, Brisbane, Australia.

Grantz, D.A. (1990) Plant-response to atmospheric humidity. *Plant Cell and Environment* 13, 667–679.

Gregg, P.C. and Wilson, L.J. (2008) The changing climate for entomology. *14th Australian Cotton Conference, Cotton Australia*, Broadbeach, Queensland, Australia.

Guinn, G. (1974) Abscission of cotton floral buds and bolls as influenced by factors affecting photosynthesis and respiration. *Crop Science* 14, 291–293.

Guinn, G. (1985) Fruiting of cotton. III. Nutritional stress and cutout. *Crop Science* 25, 981–985.

Guinn, G. and Brummett, D.L. (1989) Fruiting of cotton. IV. Nitrogen, abscisic acid, indole-3-acetic acid, and cutout. *Field Crops Research* 22, 257–266.

Guo, W.Q., Liu, R.X., Zhou, Z.G., Chen, B.L. and Xu, N.Y. (2010) Waterlogging of cotton calls for caution with N fertilization. *Acta Agriculturae Scandinavica Section B, Soil and Plant Science* 60, 450–459.

Guo, X.B., Drury, C.F., Yang, X.M., Reynolds, W.D. and Fan, R.Q. (2014) The extent of soil drying and rewetting affects nitrous oxide emissions, denitrification, and nitrogen mineralization. *Soil Science Society of America Journal* 78, 194–204.

Hall, A.E. (2001) *Crop Responses to Environment.* CRC Press, Boca Raton, Florida.

Hammer, G. (2000) A general systems approach to applying seasonal climate forecasts. In: Hammer, G.L., Nicholls, N. and Mitchell, C. (eds) *Applications of Seasonal Climate Forecasting in Agricultural and Natural Ecosystems: The Australian Experience.* Springer, Dordrecht, p. 51.

Hammer, G.L. and Nicholls, N. (1996) Managing for climate variability – the role of seasonal climate forecasting in improving agricultural systems. *Second Australian Conference on Agricultural Meteorology*, Bureau of Meteorology, Commonwealth of Australia, Melbourne, Australia, pp. 19–27.

Hammer, G.L. and Wright, G.C. (1994) A theoretical analysis of nitrogen and radiation effects on radiation use efficiency in peanuts. *Australian Journal of Agricultural Research* 45, 575–589.

Hanstein, S.M. and Felle, H.H. (2002) CO_2-triggered chloride release from guard cells in intact Fava bean leaves. Kinetics of the onset of stomatal closure. *Plant Physiology* 130, 940–950.

Hardaker, J.B., Huirne, R.B.M. and Anderson, J.R. (1997) *Coping with Risk in Agriculture.* CAB International, Wallingford, UK.

Harley, P.C., Thomas, R.B., Reynolds, J.F. and Strain, B.R. (1992) Modelling photosynthesis of cotton grown in elevated CO_2. *Plant, Cell and Environment* 15, 271–282.

Harris, R.H., Officer, S.J., Hill, P.A., Armstrong, R.D., Fogarty, K.M., Zollinger, R.P., Phelan, A.J. and Partington, D.L. (2013) Can nitrogen fertiliser and nitrification inhibitor management influence N_2O losses from high rainfall cropping systems in South Eastern Australia? *Nutrient Cycling in Agroecosystems* 95, 269–285.

Hartmann, A.A., Barnard, R.L., Marhan, S. and Niklaus, P.A. (2013) Effects of drought and N-fertilization on N cycling in two grassland soils. *Oecologia* 171, 705–717.

Hatfield, J.L. and Walthall, C.L. (2015) Meeting global food needs: realizing the potential via genetics × environment × management interactions. *Agronomy Journal* 107, 1215–1226.

Hatfield, J.L., Boote, K.J., Kimball, B.A., Ziska, L.H., Izaurralde, R.C., Ort, D., Thomson, A.M. and Wolfe, D. (2011) Climate impacts on agriculture: implications for crop production. *Agronomy Journal* 103, 351–370.

He, Z.L., Xiong, J.B., Kent, A.D., Deng, Y., Xue, K., Wang, G.J., Wu, L.Y., Van Nostrand, J.D. and Zhou, J.Z. (2014) Distinct responses of soil microbial communities to elevated CO_2 and O_3 in a soybean agro-ecosystem. *ISME Journal* 8, 714–726.

Heagle, A.S. (2003) Influence of elevated carbon dioxide on interactions between *Frankliniella occidentalis* and *Trifolium repens*. *Environmental Entomology* 32, 421–424.

Hearn, A.B. (1969) The growth and performance of cotton in a desert environment. II. Dry matter production. *Journal of Agricultural Science, Cambridge* 73, 75–86.

Hearn, A.B. (1972) The growth and performance of rain-grown cotton in a tropical upland environment. II. The relationship between yield and growth. *Journal of Agricultural Science* 79, 137–145.

Hearn, A.B. (1976a) Crop physiology. In: Arnold, M.H. (ed.) *Agricultural Research for Development*. Cambridge University Press, Cambridge, UK, pp. 77–122.

Hearn A.B. (1976b) Response of cotton to nitrogen and water in a tropical environment III. Fibre quality. *Journal of Agricultural Science, Cambridge* 86, 257–269.

Hearn, A.B. (1980) Water relationships in cotton. *Outlook on Agriculture* 10, 159–166.

Hearn, A.B. (1992) Risk and reduced water allocations. *The Australian Cotton Grower* 13(5), 50–55.

Hearn, A.B. (1994) OZCOT: A simulation model for cotton crop management. *Agricultural Systems* 44, 257–299.

Hearn, A.B. and Bange, M.P. (2002) SIRATAC and CottonLOGIC: persevering with DSSs in the Australian cotton industry. *Agricultural Systems* 74, 27–56.

Hearn, A.B. and Constable, G.A. (1984) Cotton. In: Goldsworthy, P.R. and Fisher, N.M. (eds) *The Physiology of Tropical Field Crops*. Wiley, Chichester, UK, pp. 495–527.

Hendrix, D.L., Mauney, J.R., Kimball, B.A., Lewin, K., Nagy, J. and Hendrey, G.R. (1994) Influence of elevated CO_2 and mild water stress on nonstructural carbohydrates in field-grown cotton tissues. *Agricultural and Forest Meteorology* 70, 153–162.

Hesketh, J.D. and Low, A. (1968) Effect of temperature on components of yield and fibre quality of cotton varieties of diverse origin. *Cotton Growing Review* 45, 243–257.

Hesketh, J.D., Baker, D.N. and Duncan, W.G. (1972) Simulation of growth and yield in cotton. 2. Environmental control of morphogenesis. *Crop Science* 12, 436–439.

Hileman, D.R., Huluka, G., Kenjige, P.K., Sinha, N., Bhattacharya, N.C., Biswas, P.K., Lewin, K.F., Nagy, J. and Hendrey, G.R. (1994) Canopy photosynthesis and transpiration of field-grown cotton exposed to free-air CO_2 enrichment (FACE) and differential irrigation. *Agricultural and Forest Meteorology* 70, 189–207.

Hochman, Z., Gobbett, D., Holzworth, D., McClelland, T., van Rees, H., Marinoni, O., Garcia, J.N. and Horan, H. (2012) Quantifying yield gaps in rainfed cropping systems: A case study of wheat in Australia. *Field Crops Research* 136, 85–96.

Hodges, H.F., Reddy, K.R., McKinion, J.M. and Reddy, V.R. (1993) Temperature effects on cotton. *Bulletin* 990, Department of Information Services, Division of Agriculture, Forestry and Veterinary Medicine, Mississippi State University.

Hodgson, A.S. (1982) The effects of duration, timing and chemical amerlioration of short-term waterlogging in a cracking grey clay. *Australian Journal of Agricultural Research* 33, 1019–1028.

Hodgson, A.S. and Chan, K.Y. (1982) The effect of short-term waterlogging during furrow irrigation of cotton in a cracking grey clay. *Australian Journal of Agricultural Resources* 33, 109–116.

Hodgson, A.S. and Macleod, D.A. (1987) Effects of foliar applied nitrogen fertilizer on cotton waterlogged in a cracking grey clay. *Australian Journal of Agricultural Research* 38, 681–688.

Hoekstra, A.Y. and Hung, P.Q. (2002) Virtual water trade: A quantification of virtual water flows between nations in relation to international crop trade. *Value of Water Research Report Series* No. 11, UNESCO-IHE Institute for Water Education, Delft, the Netherlands.

Hofs, J.L., Hau, B. and Marais, D. (2006) Boll distribution patterns in Bt and non-Bt cotton cultivars. I. Study on commercial irrigated farming systems in South Africa. *Field Crops Research* 98, 203–209.

Holtum, J.A.M. and Winter, K. (2003) Photosynthetic CO_2 uptake in seedlings of two tropical tree species exposed to oscillating elevated concentrations of CO_2. *Planta* 218, 152–158.

Hoque, Z., Farquharson, B., Dillon, M. and Kauter, G. (2000) Soft options can reduce costs and increase cotton profits. *The Australian Cottongrower* 21, 33–37.

Howden, S.M., Gifford, R.G. and Meinke, H. (2010) GRAINS. In: Stokes, C. and Howden, M. (eds) *Adapting Agriculture to Climate Change: Preparing Australian Agriculture, Forestry and Fisheries for the Future*. CSIRO Publishing, Collingwood, Victoria, Australia, pp. 21–48.

Hu, S.J., Tu, C., Chen, X. and Gruver, J.B. (2006) Progressive N limitation of plant response to elevated CO_2: a microbiological perspective. *Plant and Soil* 289, 47–58.

Huck, M.G. (1970) Variation in taproot elongation rate as influenced by composition of the soil air. *Agronomy Journal* 62, 815–818.

Hulugalle, N.R., Weaver, T.B. and Ghadiri, H. (2005a) A simple method to estimate the value of salt land nutrient leaching in irrigated vertisols in Australia. *Sustainable Use and Management of Soils – Arid and Semiarid Regions* 36, 579–588.

Hulugalle, N.R., Weaver, T.B. and Scott, F. (2005b) Continuous cotton and cotton-wheat rotation effects on soil properties and profitability in an irrigated vertisol. *Journal of Sustainable Agriculture* 27, 5–24.

Hulugalle, N.R., Weaver, T.B., Finlay, L.A., Hare, J. and Entwistle, P.C. (2007) Soil properties and crop yields in a dryland vertisol sown with cotton-based crop rotations. *Soil and Tillage Research* 93, 356–369.

Hulugalle, N.R., Weaver, T.B., Finlay, L.A. and Heimoana, V. (2013) Soil organic carbon concentrations and storage in irrigated cotton cropping systems sown on permanent beds in a vertosol with restricted subsoil drainage. *Crop and Pasture Science* 64, 799–805.

Huluka, G., Hileman, D.R., Biswas, P.K., Lewin, K.F., Nagy, J. and Hendrey, G.R. (1994) Effects of elevated CO_2 and water stress on mineral concentration of cotton. *Agricultural and Forest Meteorology* 70, 141–152.

51

Humbert, S., Tarnawski, S., Fromin, N., Mallet, M.P., Aragno, M. and Zopfi, J. (2010) Molecular detection of anammox bacteria in terrestrial ecosystems: distribution and diversity. *ISME Journal* 4, 450–454.

Hungate, B.A., Chapin, F.S., Zhong, H., Holland, E.A. and Field, C.B. (1997a) Stimulation of grassland nitrogen cycling under carbon dioxide enrichment. *Oecologia* 109, 149–153.

Hungate, B.A., Holland, E.A., Jackson, R.B., Chapin, F.S., Mooney, H.A. and Field, C.B. (1997b) The fate of carbon in grasslands under carbon dioxide enrichment. *Nature* 388, 576–579.

Hunsaker, D.J., Hendrey, G.R., Kimball, B.A., Lewin, K.F., Mauney, J.R. and Nagy, J. (1994) Cotton evapotranspiration under field conditions with CO_2 enrichment and variable soil moisture regimes. *Agricultural and Forest Meteorology* 70, 247–258.

Idso, S.B., Kimball, B.A., Wall, G.W., Garcia, R.L., LaMorte, R., Pinter, P.J., Jr, Mauney, J.R., Hendrey, G.R., Lewin, K. and Nagy, J. (1994) Effects of free-air CO_2 enrichment on the light response curve of net photosynthesis in cotton leaves. *Agricultural and Forest Meteorology* 70, 183–188.

IPCC (2007a) Climate Change 2007: Synthesis Report. In: Pachauri, R.K. and Reisinger, A. (eds) *Contribution of Working Groups I, II and III to the Fourth Assessment Report of the Intergovernmental Panel on Climate Change*. IPCC, Geneva.

IPCC (2007b) Climate Change 2007: The Physical Science Basis. In: Solomon, S., Qin, D., Manning, M., Chen, Z., Marquis, M., Averyt, K.B., Tignor, M. and Miller, H.L. (eds) *Contribution of Working Group I to the Fourth Assessment Report of the Intergovernmental Panel on Climate Change*. Cambridge University Press, Cambridge, UK.

IPCC (2013) Climate Change 2013: The Physical Science Basis. Available at: http://www.ipcc.ch/report/ar5/wg1 (accessed 24 November 2015).

IPCC (2014) Climate Change 2014: Synthesis Report. In: Pachauri, R.K. and Meyer, L.A. (eds) *Contribution of Working Groups I, II and III to the Fifth Assessment Report of the Intergovernmental Panel on Climate Change*. IPCC, Geneva, p. 151.

ITC (2011) *Cotton and Climate Change – Impacts and Options to Mitigate and Adapt*. International Trade Centre, Geneva, 32 pp.

Jackson, L.E., Burger, M. and Cavagnaro, T.R. (2008) Roots, nitrogen transformations, and ecosystem services. *Plant Biology* 59, 341.

Jackson, M.B. (1985) Ethylene and responses of plants to soil waterlogging and submergence. *Annual Review of Plant Physiology* 36, 145–174.

Jackson, M.B. and Drew, M.C. (1984) Effects of flooding on growth and metabolism of herbaceous plants. In: Kozlowski, T.T. (ed.) *Flooding and Plant Growth*. Academic Press, New York, pp. 47–128.

Jackson, R.B., Cook, C.W., Pippen, J.S. and Palmer, S.M. (2009) Increased belowground biomass and soil CO_2 fluxes after a decade of carbon dioxide enrichment in a warm-temperate forest. *Ecology* 90, 3352–3366.

Jagadish, K.S.V., Kadam, N.N., Xiao, G., Melgar, R.J., Bahuguna, R.N., Quinones, C., Tamilselvan, A. and Prasad, P.V.V. (2014) Agronomic and physiological responses to high temperature, drought, and elevated CO_2 interactions in cereals. *Advances in Agronomy* 127, 111–156.

Jahnke, S., Krewitt, M., Pieruschka, R. and Kohnen, M. (2001) Leaf respiration is not directly affected by elevated CO_2 concentration: Facts and artefacts in respiration analysis. In: *PS 2001 Proceedings: 12th International Congress on Photosynthesis*, Brisbane, Queensland, Australia, 18–23 August, 2001. CSIRO Publishing, Melbourne, Australia.

Jensen, E.S., Peoples, M.B., Boddey, R.M., Gresshoff, P.M., Hauggaard-Nielsen, H., Alves, B.J.R. and Morrison, M.J. (2012) Legumes for mitigation of climate change and the provision of feedstock for biofuels and biorefineries. A review. *Agronomy for Sustainable Development* 32, 329–364.

Jordon, W.R. and Ritchie, J.T. (1971) Influence of soil water stress on evaporation, root absorption, and internal water status of cotton. *Plant Physiology* 48, 783–788.

Kaisermann, A., Roguet, A., Nunan, N., Maron, P.A., Ostle, N. and Lata, J.C. (2013) Agricultural management affects the response of soil bacterial community structure and respiration to water-stress. *Soil Biology and Biochemistry* 66, 69–77.

Kakani, V.G., Reddy, K.R., Zhao, D. and Gao, W. (2004) Senescence and hyperspectral reflectance of cotton leaves exposed to ultraviolet-B radiation and carbon dioxide. *Physiologia Plantarum* 121, 250–257.

Kakani, V.G., Reddy, K.R., Koti, S., Wallace, T.P., Prasad, P.V.V., Reddy, V.R. and Zhao, D. (2005) Differences in *in vitro* pollen germination and pollen tube growth of cotton cultivars in response to high temperature. *Annals of Botany* 96, 59–67.

Kawakami, E., Oosterhuis, D. and Snider, J. (2010) Physiological effects of 1-methylcyclopropene on well-watered and water-stressed cotton plants. *Journal of Plant Growth Regulation* 29, 280–288.

Kawakami, E.M., Oosterhuis, D.M., Snider, J.L. and Mozaffari, M. (2012) Physiological and yield responses of field-grown cotton to application of urea with the urease inhibitor NBPT and the nitrification inhibitor DCD. *European Journal of Agronomy* 43, 147–154.

Keil, D., Niklaus, P.A., von Riedmatten, L.R., Boeddinghaus, R.S., Dormann, C.F., Scherer-Lorenzen, M., Kandeler, E. and Marhan, S. (2015) Effects of warming and drought on potential N_2O emissions and denitrifying bacteria abundance in grasslands with different land-use. *FEMS Microbiology Ecology* 91(7). DOI: 10.1093/femsec/fiv066.

Kim, S.-H., Reddy, V.R., Baker, J.T., Gitz, D.C. and Timlin, D.J. (2004) Quantification of photosynthetically active radiation inside sunlit growth chambers. *Agricultural and Forest Meteorology* 126, 117–127.

Kimball, B.A. (1983) Carbon-dioxide and agricultural yield – an assemblage and analysis of 430 prior observations. *Agronomy Journal* 75, 779–788.

Kimball, B.A. (2005) Theory and performance of an infrared heater for ecosystem warming. *Global Change Biology* 11, 2041–2056.

Kimball, B.A. and Idso, S.B. (1983) Increasing atmospheric CO_2: effects on crop yield, water use and climate. *Agricultural Water Management* 7, 55–72.

Kimball, B.A. and Mauney, J.R. (1993) Response of cotton to varying CO_2, irrigation and nitrogen – yield and growth. *Agronomy Journal* 85, 706–712.

Kimball, B.A., Mauney, J.R., Guinn, G., Nakayama, F.S., Pinter, P.J.J., Clawson, K.L., Idso, S., Butler, G.D. and Radin, J.W. (1984) Effects of increasing atmospheric CO_2 on the yield and water use of crops. *Response of Vegetation to Carbon Dioxide* No. 023. US Departments of Energy and Agriculture, Washington.

Kimball, B.A., Mauney, J.R., Guinn, G., Nakayama, F.S., Idso, S., Radin, J.W., Hendrix, D.L., Butler, G.D., Zarembinski, T.J. and Nixon, P.E. (1985) Effects of increasing atmospheric CO_2 on the yield and water use of crops. *Response of Vegetation to Carbon Dioxide* No. 027. US Departments of Energy and Agriculture, Washington.

Kimball, B.A., LaMorte, R.L., Seay, R.S., Pinter, P.J., Jr, Rokey, R.R., Hunsaker, D.J., Dugas, W.A., Heuer, M.L., Mauney, J.R., Hendrey, G.R., Lewin, K.F. and Nagy, J. (1994) Effects of free-air CO_2 enrichment on energy balance and evapotranspiration of cotton. *Agricultural and Forest Meteorology* 70, 259–278.

Kimball, B.A., Pinter, P.J., Wall, G.W., Jr, Garcia, R.L., LaMorte, R.L., Jak, P.M.C., Frumau, K.F.A. and Vugts, H.F. (1997) Comparisons of responses of vegetation to elevated carbon dioxide in free-air and open-top chamber facilities. In: Allen, L.H., Kirkham, M.B., Olszyk, D.M. and Whitman, C.E. (eds) *Advances in Carbon Dioxide Effects Research*. ASA/CSSA/SSSA, Madison, Wisconsin, pp. 113–130.

Kimball, B.A., White, J.W., Wall, G.W. and Ottman, M.J. (2012) Infrared-warmed and unwarmed wheat vegetation indices coalesce using canopy-temperature–based growing degree days. *Agronomy Journal* 104, 114–118.

Knox, B.R., Ladiges, P.Y., Evans, B.K. and Saint, R. (2005) *Biology: an Australian focus*, 3rd edn. McGraw-Hill Education, North Ryde, Australia.

Ko, J.H. and Piccinni, G. (2009) Characterizing leaf gas exchange responses of cotton to full and limited irrigation conditions. *Field Crops Research* 112, 77–89.

Korner, C. and Arnone, J.A. (1992) Responses to elevated carbon-dioxide in artificial tropical ecosystems. *Science* 257, 1672–1675.

Kranthi, K.R. (2009) Challenges and opportunities in cotton production research. In: ICAC (2009) *Biosafety Regulations, Implementation and Consumer Acceptance*. International Cotton Advisory Committee (ICAC), Washington, DC, pp. 16–20.

Krieg, D.R. and Kerby, T.A. (1985) Cotton response to mepiquat chloride. *Agronomy Journal* 77, 907–912.

Ku, S.B. and Edwards, G.E. (1977) Oxygen inhibition of photosynthesis 1. Temperature-dependence and relation to O_2-CO_2 solubility ratio. *Plant Physiology* 59, 986–990.

Lal, R., Delgado, J.A., Groffman, P.M., Millar, N., Dell, C. and Rotz, A. (2011) Management to mitigate and adapt to climate change. *Journal of Soil and Water Conservation* 66, 276–285.

Langdale, G.W., Wilson, R.L. and Bruce, R.R. (1990) Cropping frequencies to sustain long-term conservation tillage systems. *Soil Science Society of America Journal* 54, 193–198.

Langley, J.A., McKinley, D.C., Wolf, A.A., Hungate, B.A., Drake, B.G. and Megonigal, J.P. (2009) Priming depletes soil carbon and releases nitrogen in a scrub-oak ecosystem exposed to elevated CO_2. *Soil Biology & Biochemistry* 41, 54–60.

Lascano, R.J., Baumhardt, R.L., Hicks, S.K. and Heilman, J.L. (1994) Soil and plant water evaporation from strip-tilled cotton – measurement and simulation. *Agronomy Journal* 86, 987–994.

Le Houérou, H.N. (1996) Climate change, drought and desertification. *Journal of Arid Environments* 34, 133–185.

Leakey, A.D.B., Xu, F., Gillespie, K.M., McGrath, J.M., Ainsworth, E.A. and Ort, D.R. (2009) Genomic basis for stimulated respiration by plants growing under elevated carbon dioxide. *Proceedings of the National Academy of Sciences of the United States of America* 106, 3597–3602.

Leonard, O.A. and Pinckard, J.A. (1946) Effect of various oxygen and carbon dioxide concentrations on cotton root development. *Plant Physiology* 21, 18–36.

Liakatas, A., Roussopoulos, D. and Whittington, W.J. (1998) Controlled-temperature effects on cotton yield and fibre properties. *Journal of Agricultural Science, Cambridge* 130, 463–471.

Liang, X.Z., Xu, M., Gao, W., Reddy, K.R., Kunkel, K., Schmoldt, D.L. and Samel, A.N. (2012a) A distributed cotton growth model developed from GOSSYM and its parameter determination. *Agronomy Journal* 104, 661–674.

Liang, X.Z., Xu, M., Gao, W., Reddy, K.R., Kunkel, K., Schmoldt, D.L. and Samel, A.N. (2012b) Physical modeling of US cotton yields and climate stresses during 1979 to 2005. *Agronomy Journal* 104, 675–683.

Liu, C.-A., Zhou, L.-M., Jia, J.-J., Wang, L.-J., Si, J.-T., Li, X., Pan, C.-C., Siddique, K.H.M. and Li, F.-M. (2014) Maize yield and water balance is affected by nitrogen application in a film-mulching ridge–furrow system in a semiarid region of China. *European Journal of Agronomy* 52(Part B), 103–111.

Liu, P., Sun, F., Gao, R. and Dong, H. (2012) RAP2.6L overexpression delays waterlogging induced premature senescence by increasing stomatal closure more than antioxidant enzyme activity. *Plant Molecular Biology* 79, 609–622.

Liu, Q., Singh, S.P. and Green, A.G. (2002) High-stearic and high-oleic cottonseed oils produced by hairpin RNA-mediated post-transcriptional gene silencing. *Plant Physiology* 129, 1732–1743.

Liu, S.M., Constable, G.A., Reid, P.E., Stiller, W.N. and Cullis, B.R. (2013) The interaction between breeding and crop management in improved cotton yield. *Field Crops Research* 148, 49–60.

Loka, D.A. and Oosterhuis, D.M. (2010) Effect of high night temperatures on cotton respiration, ATP levels and carbohydrate content. *Environmental and Experimental Botany* 68, 258–263.

Loka, D. and Oosterhuis, D. (2013) Effect of 1-MCP on gas exchange and carbohydrate concentrations of the cotton flower and subtending leaf under water-deficit stress. *American Journal of Plant Sciences* 4, 142–152.

Loka, D.A., Oosterhuis, D.M. and Ritchie, G.L. (2011) Water-deficit stress in cotton. In: Oosterhuis, D.M. (ed.) *Stress Physiology in Cotton*. The Cotton Foundation, Cordova, Tennessee, pp. 37–72.

Lokhande, S. and Reddy, K.R. (2014a) Quantifying temperature effects on cotton reproductive efficiency and fiber quality. *Agronomy Journal* 106, 1275–1282.

Lokhande, S. and Reddy, K.R. (2014b) Reproductive and fiber quality responses of upland cotton to moisture deficiency. *Agronomy Journal* 106, 1060–1069.

Lokhande, S. and Reddy, K.R. (2015a) Reproductive performance and fiber quality responses of cotton to potassium nutrition. *American Journal of Plant Sciences* 6, 911–924.

Lokhande, S.B. and Reddy, K.R. (2015b) Cotton reproductive and fiber quality responses to nitrogen nutrition. *International Journal of Plant Production* 9, 191–209.

Long, A., Heitman, J., Tobias, C., Philips, R. and Song, B. (2013) Co-occurring anammox, denitrification, and codenitrification in agricultural soils. *Applied and Environmental Microbiology* 79, 168–176.

Long, S., Ainsworth, E.A., Leakey, A.D.B., Nosberger, J. and Ort, D.R. (2006) Food for thought: lower-than-expected crop yield stimulation with rising CO_2 concentrations. *Science (New York)* 312, 1918–1921.

Lu, Z., Chen, J., Percy, R.G. and Zeiger, E. (1997) Photosynthetic rate, stomatal conductance and leaf area in two cotton species (*Gossypium barbadense* and *Gossypium hirsutum*) and their relation with heat resistance and yield. *Australian Journal of Plant Physiology* 24(5), 693–700.

Lu, Z.M., Percy, R.G., Qualset, C.O. and Zeiger, E. (1998) Stomatal conductance predicts yields in irrigated Pima cotton and bread wheat grown at high temperatures. *Journal of Experimental Botany* 49, 453–460.

Luo, Q.Y., Bange, M. and Clancy, L. (2014) Cotton crop phenology in a new temperature regime. *Ecological Modelling* 285, 22–29.

Luo, Q.Y., Bange, M., Johnston, D. and Braunack, M. (2015) Cotton crop water use and water use efficiency in a changing climate. *Agriculture Ecosystems & Environment* 202, 126–134.

Luo, Y., Su, B., Currie, W.S., Dukes, J.S., Finzi, A., Hartwig, U., Hungate, B., McMurtrie, R.E., Oren, R., Parton, W.J., Pataki, D.E., Shaw, M.R., Zak, D.R. and Field, C.B. (2004) Progressive nitrogen limitation of ecosystem responses to rising atmospheric carbon dioxide. *Bioscience* 54, 731–739.

Luo, Y.Q., Hui, D.F. and Zhang, D.Q. (2006) Elevated CO_2 stimulates net accumulations of carbon and nitrogen in land ecosystems: a meta-analysis. *Ecology* 87, 53–63.

Macdonald, C. and Singh, B. (2014) Harnessing plant-microbe interactions for enhancing farm productivity. *Bioengineered* 5, 5–9.

Macrobbie, E.A.C. (1983) Effects of light/dark on cation fluxes in guard cells of *Commelina communis* L. *Journal of Experimental Botany* 34, 1695–1710.

Manderscheid, R. and Weigel, H.J. (1997) Photosynthetic and growth responses of old and modern spring wheat cultivars to atmospheric CO_2 enrichment. *Agriculture Ecosystems & Environment* 64, 65–73.

Mansoor, S. and Paterson, A.H. (2012) Genomes for jeans: cotton genomics for engineering superior fiber. *Trends in Biotechnology* 30, 521–527.

Marani, A. and Amirav, A. (1971) Effects of soil moisture stress on two cultivars of upland cotton in Israel: I. The coastal plain region. *Experimental Agriculture* 7, 213–224.

Maraseni, T.N., Cockfield, G. and Maroulis, J. (2010) An assessment of greenhouse gas emissions: implications for the Australian cotton industry. *Journal of Agricultural Science* 148, 501–510.

Mason, T.G. (1922) Growth and abscission in Sea Island cotton. *Annals of Botany* 36, 458–484.

Massacci, A., Nabiev, S.M., Pietrosanti, L., Nematov, S.K., Chernikova, T.N., Thor, K. and Leipner, J. (2008) Response of the photosynthetic apparatus of cotton (*Gossypium hirsutum*) to the onset of drought stress under field conditions studied by gas-exchange analysis and chlorophyll fluorescence imaging. *Plant Physiology and Biochemistry* 46, 189–195.

Mauney, J.R., Fry, K.E. and Guinn, G. (1978) Relationship of photosynthetic rate to growth and fruiting of cotton, soybean, sorghum and sunflower. *Crop Science* 18, 259–263.

Mauney, J.R., Kimball, B.A., Pinter, P.J., Jr, LaMorte, R.L., Lewin, K.F., Nagy, J. and Hendrey, G.R. (1994) Growth and yield of cotton in response to a free-air carbon dioxide enrichment (FACE) environment. *Agricultural and Forest Meteorology* 70, 49–67.

Maurino, V.G. and Weber, A.P.M. (2013) Engineering photosynthesis in plants and synthetic microorganisms. *Journal of Experimental Botany* 64, 743–751.

McCarthy, A.C., Hancock, N.H. and Raine, S.R. (2010) VARIwise: A general-purpose adaptive control simulation framework for spatially and temporally varied irrigation at sub-field scale. *Computers and Electronics in Agriculture* 70, 117–128.

McGrath, J.M. and Long, S.P. (2014) Can the cyanobacterial carbon-concentrating mechanism increase photosynthesis in crop species? A theoretical analysis. *Plant Physiology* 164, 2247–2261.

McIntosh, P.C., Asseng, S. and Wang, E. (2015) Profit and risk in dryland cropping: seasonal forecasts and fertiliser management. *17th Australian Agronomy Conference*, Australian Society of Agronomy, Hobart, Tasmania, Australia.

McLeod, A.R. and Long, S.P. (1999) Free-air carbon dioxide enrichment (FACE) in global change research: A review. *Advances in Ecological Research* 28, 1–56.

McMichael, B.L. and Burke, J.J. (1994) Metabolic activity of cotton roots in response to temperature. *Environmental and Experimental Botany* 34, 201–206.

Medrano, H., Escalona, J.M., Bota, J., Gulias, J. and Flexas, J. (2002) Regulation of photosynthesis of C_3 plants in response to progressive drought: Stomatal conductance as a reference parameter. *Annals of Botany* 89, 895–905.

Meisner, A., Rousk, J. and Baath, E. (2015) Prolonged drought changes the bacterial growth response to rewetting. *Soil Biology & Biochemistry* 88, 314–322.

Meyer, V.G. (1966) Environmental effects on the differentiation of abnormal cotton flowers. *American Journal of Botany* 53, 976–980.

Meyer, W.S., Reicosky, D.C., Barrs, H.D. and Smith, R.C.G. (1987) Physiological responses of cotton to a single waterlogging at high and low N-levels. *Plant and Soil* 102, 161–170.

Miglietta, F., Raschi, A., Bettarini, I., Resti, R. and Selvi, F. (1993) Natural CO_2 springs in Italy – a resource for examining long-term response of vegetation to rising atmospheric CO_2 concentrations. *Plant Cell and Environment* 16, 873–878.

Milroy, S.P. and Bange, M.P. (2013) Reduction in radiation use efficiency of cotton (*Gossypium hirsutum* L.) under repeated transient waterlogging in the field. *Field Crops Research* 140, 51–58.

Milroy, S.P., Bange, M.P. and Thongbai, P. (2009) Cotton leaf nutrient concentrations in response to waterlogging under field conditions. *Field Crops Research* 113, 246–255.

Montgomery, J. and O'Halloran, J. (2008) A comparison of water use between solid plant and one–in–out skip. *The Australian Cottongrower* 29, 21–25.

Morison, J.I.L. and Gifford, R.M. (1983) Stomatal sensitivity to carbon-dioxide and humidity – a comparison of 2 C_3 and 2 C_4 grass species. *Plant Physiology* 71, 789–796.

Morison, J.I.L. and Gifford, R.M. (1984) Plant-growth and water-use with limited water-supply in high CO_2 concentrations. 1. Leaf-area, water-use and transpiration. *Australian Journal of Plant Physiology* 11, 361–374.

Moss, J., Gordon, I. and Zischke, R. (1999) Vertosols do 'leak' – water and solute movement below irrigated cotton. In: *Proceedings of the Murray Darling Basin Groundwater Workshop 1999*, MDBC, Griffith, New South Wales, Australia, pp. 298–303.

Mott, K.A. (1988) Do stomata respond to CO_2 concentrations other than intercellular? *Plant Physiology (Bethesda)* 86, 200–203.

Moya, T.B., Ziska, L.H., Namuco, O.S. and Olszyk, D. (1998) Growth dynamics and genotypic variation in tropical, field-grown paddy rice (*Oryza sativa* L.) in response to increasing carbon dioxide and temperature. *Global Change Biology* 4, 645–656.

Muller, C., Rutting, T., Abbasi, M.K., Laughlin, R.J., Kammann, C., Clough, T.J., Sherlock, R.R., Kattge, J., Jager, H.J., Watson, C.J. and Stevens, R.J. (2009) Effect of elevated CO_2 on soil N dynamics in a temperate grassland soil. *Soil Biology & Biochemistry* 41, 1996–2001.

Najeeb, U., Atwell, B., Bange, M. and Tan, D.Y. (2015a) Aminoethoxyvinylglycine (AVG) ameliorates waterlogging-induced damage in cotton by inhibiting ethylene synthesis and sustaining photosynthetic capacity. *Plant Growth Regulation* 76, 83–98.

Najeeb, U., Tan, D.K.Y. and Bange, M.P. (2015b) Inducing waterlogging tolerance in cotton via an anti-ethylene agent aminoethoxyvinylglycine application. *Archives of Agronomy and Soil Science*, DOI: 10.1080/03650340.2015.1113403.

Nakayama, F.S., Huluka, G., Kimball, B.A., Lewin, K.F., Nagy, J. and Hendrey, G.R. (1994) Soil carbon dioxide fluxes in natural and CO_2-enriched systems. *Agricultural and Forest Meteorology* 70, 131–140.

Nakicenovic, N. and Swart, R. (2000) *Emissions Scenarios. Special report of the Intergovernmental Panel on Climate Change*. Cambridge University Press, Cambridge, UK.

NCAR (2001) *A Theoretical Analysis of the Effect of the Spatial Resolution of Climate Scenarios on Cotton Production in the Southeastern United States*. National Center for Atmospheric Research, Boulder, Colorado.

NCAR (2004) *Climate Change Impacts*. Brochure, National Center for Atmospheric Research, Boulder, Colorado.

Norby, R.J., Ledford, J., Reilly, C.D., Miller, N.E. and O'Neill, E.G. (2004) Fine-root production dominates response of a deciduous forest to atmospheric CO_2 enrichment. *Proceedings of the National Academy of Sciences of the United States of America* 101, 9689–9693.

Oosterhuis, D.M. (1999) Yield response to environmental extremes in cotton. In: Oosterhuis, D.M. (ed.) *Cotton Research Meeting Summary, Cotton Research in Progress*. Arkansas Agriculture Experiment Station, Fayetteville, Arkansas, pp. 30–38.

Oosterhuis, D.M. (2002) Day or night high temperature: A major cause of yield variability. *Cotton Grower* 46, 8–9.

Oosterhuis, D.M. (2013) Global warming and cotton productivity. In: ICA Committee (ed.) *International Cotton Advisory Committee 72nd Plenary Meeting*, Cartagena, Colombia.

Oosterhuis, D.M. and Brown, R.S. (2005) Increased nitrogen and protein content with the growth regulator Chaperone™. In: Li, C.J. (ed.) *Plant Nutrition for Food Security, Human Health, and Environmental Protection*. Tsinghua University Press, Beijing, China, pp. 1158–1159.

Oosterhuis, D.M. and Jernstedt, J. (1999) Morphology and anatomy of the cotton plant. In: Smith, C.W. and Cothren, J.T. (eds) *Cotton Origin, History, Technology, and Production*. John Wiley and Sons, New York, pp. 175–206.

Oosterhuis, D.M. and Snider, J.L. (2011) High temperature stress on floral development and yield of cotton. In: Oosterhuis, D.M. (ed.) *Stress Physiology in Cotton*. Cotton Foundation, Cordova, Tennessee, pp. 1–12.

Oren, R., Sperry, J.S., Katul, G.G., Pataki, D.E., Ewers, B.E., Phillips, N. and Schafer, K.V.R. (1999) Survey and synthesis of intra- and interspecific variation in stomatal sensitivity to vapour pressure deficit. *Plant Cell and Environment* 22, 1515–1526.

Pakistan (2003) Pakistan's Initial National Communication on Climate Change. Ministry of Environment, Islamabad, Pakistan, 92 pp.

Palle, S.R., Campbell, L.M., Pandeya, D., Puckhaber, L., Tollack, L.K., Marcel, S., Sundaram, S., Stipanovic, R.D., Wedegaertner, T.C., Hinze, L. and Rathore, K.S. (2013) RNAi-mediated Ultra-low gossypol cottonseed trait: performance of transgenic lines under field conditions. *Plant Biotechnology Journal* 11, 296–304.

Pandey, D.M., Goswami, C.L., Kumar, B. and Jain, S. (2001) Hormonal regulation of photosynthetic enzymes in cotton under water stress. *Photosynthetica* 38, 403–407.

Paterson, A.H., Saranga, Y., Menz, M., Jiang, C.X. and Wright, R.J. (2003) QTL analysis of genotype × environment interactions affecting cotton fiber quality. *Theoretical and Applied Genetics* 106, 384–396.

Pearson, L.C. (1967) *Principles of Agronomy*. Reinhold Publishing Corporation, New York.

Pearson, R.W., Ratliff, L.R. and Taylor, H.M. (1970) Effect of soil temperature, strength and pH on cotton seedling root elongation. *Agronomy Journal* 87, 947–952.

Pendall, E., Bridgham, S., Hanson, P.J., Hungate, B., Kicklighter, D.W., Johnson, D.W., Law, B.E., Luo, Y.Q., Megonigal, J.P., Olsrud, M., Ryan, M.G. and Wan, S.Q. (2004) Below-ground process responses to elevated CO_2 and temperature: a discussion of observations, measurement methods, and models. *New Phytologist* 162, 311–322.

Perry, L.G., Shafroth, P.B., Blumenthal, D.M., Morgan, J.A. and LeCain, D.R. (2013) Elevated CO_2 does not offset greater water stress predicted under climate change for native and exotic riparian plants. *New Phytologist* 197, 532–543.

Pettigrew, W.T. (1995) Source-to-sink manipulation effects on cotton fiber quality. *Agronomy Journal* 87, 947–952.

Pettigrew, W.T. (2004a) Moisture deficit effects on cotton lint yield, yield components, and boll distribution. *Agronomy Journal* 96, 377–383.

Pettigrew, W.T. (2004b) Physiological consequences of moisture deficit stress in cotton. *Crop Science* 44, 1265–1272.

Pettigrew, W.T. (2008) The effect of higher temperatures on cotton lint yield production and fibre quality. *Crop Science* 48, 278–285.

Pettigrew, W.T. and Adamczyk, J.J. (2006) Nitrogen fertility and planting date effects on lint yield and Cry1Ac (Bt) endotoxin production. *Agronomy Journal* 98, 691–697.

Pettigrew, W.T. and Oosterhuis, D.M. (2013) *Cotton, Climate Change and Agriculture: Effects and Adaptation*. National Climate Assessment for Agriculture. US Global Change Research Program, Washington, DC.

Pettigrew, W.T., Hesketh, J.D., Peters, D.B. and Woolley, J.T. (1990) A vapor pressure deficit effect on crop canopy photosynthesis. *Photosynthesis Research* 24, 27–34.

Phene, C.J., Baker, D.N., Lambert, J.R., Parsons, J.E. and McKinion, J.M. (1978) SPAR – A soil-plant-atmosphere-research system. *Transactions of the ASAE* 21, 24–30.

Phillips, R.E., Blevins, R.L., Thomas, G.W., Frye, W.W. and Phillips, S.H. (1980) No-tillage agriculture. *Science* 208, 1108–1113.

Pinter, P.J., Jr, Idso, S.B., Hendrix, D.L., Rokey, R.R., Rauschkolb, R.S., Mauney, J.R., Kimball, B.A., Hendrey, G.R., Lewin, K.F. and Nagy, J. (1994a) Effect of free-air CO_2 enrichment on the chlorophyll content of cotton leaves. *Agricultural and Forest Meteorology* 70, 163–169.

Pinter, P.J., Jr, Kimball, B.A., Mauney, J.R., Hendrey, G.R., Lewin, K.F. and Nagy, J. (1994b) Effects of free-air carbon dioxide enrichment on PAR absorption and conversion efficiency by cotton. *Agricultural and Forest Meteorology* 70, 209–230.

Power, B. and Cacho, O.J. (2014) Identifying risk-efficient strategies using stochastic frontier analysis and simulation: An application to irrigated cropping in Australia. *Agricultural Systems* 125, 23–32.

Power, B., Rodriguez, D., deVoil, P., Harris, G. and Payero, J. (2011) A multi-field bio-economic model of irrigated grain-cotton farming systems. *Field Crops Research* 124, 171–179.

Prior, S.A., Rogers, H.H., Runion, G.B. and Mauney, J.R. (1994) Effects of free-air CO_2 enrichment on cotton root growth. *Agricultural and Forest Meteorology* 70, 69–86.

Radin, J.W. and Ackerson, R.C. (1981) Water relations of cotton plants under nitrogen deficiency: III. Stomatal conductance, photosynthesis, and abscisic acid accummulation during drought. *Plant Physiology (Bethesda)* 67, 115–119.

Radin, J.W., Kimball, B.A., Hendrix, D.L. and Mauney, J.R. (1987) Photosynthesis of cotton plants exposed to elevated levels of carbon dioxide in the field. *Photosynthesis Research* 12, 191–204.

Ramey, H.H. (1986) Stress influences on fiber development. In: Mauney, J.R. and Stewart, J.M.D. (eds) *Cotton Physiology*. The Cotton Foundation, Memphis, Tennessee.

Raper, R.L., Reeves, D.W., Burmester, C.H. and Schwab, E.B. (2000) Tillage depth, tillage timing, and cover crop effects on cotton yield, soil strength, and tillage energy requirements. *Applied Engineering in Agriculture* 16, 379–385.

Raper, R.L., Schwab, E.B., Arriaga, F.J., Balkcom, K.S., Price, A.J. and Kornecki, T.S. (2011) Effects of cover crop removal on a cotton-peanut rotation. *Transactions of the ASABE* 54, 1213–1218.

Rawson, H.M. and Begg, J.E. (1977) The effect of atmospheric humidity on photosynthesis, transpiration and water use efficiency of leaves of several plant species. *Planta* 134, 5–10.

Rawson, H.M., Begg, J.E. and Woodward, R.G. (1977) The effect of atmospheric humidity on photosynthesis, transpiration and water use efficiency of leaves of several plant species. *Planta* 134, 5–10.

Reddy, A.R., Reddy, K.R. and Hodges, H.F. (1996) Mepiquat chloride (PIX)-induced changes in photosynthesis and growth of cotton. *Plant Growth Regulation* 20, 179–183.

Reddy, A.R., Reddy, K.R. and Hodges, H.F. (1998) Interactive effects of elevated carbon dioxide and growth temperature on photosynthesis in cotton leaves. *Plant Growth Regulation* 26, 33–40.

Reddy, K.N., Locke, M.A., Koger, C.H., Zablotowicz, R.M. and Krutz, L.J. (2006) Cotton and corn rotation under reduced tillage management: impacts on soil properties, weed control, yield, and net return. *Weed Science* 54, 768–774.

Reddy, K.R. and Hodges, H.F. (2000) Crop ecosystem responses to climatic change: Cotton. In: Reddy, K.R. and Hodges, H.F. (eds) *Climate Change and Global Productivity*. CAB International, Wallingford, UK, pp. 161–188.

Reddy, K.R. and Zhao, D. (2005) Interactive effects of elevated CO_2 and potassium deficiency on photosynthesis, growth and biomass partitioning. *Field Crops Research* 94, 201–213.

Reddy, K.R., Hodges, H.F. and Reddy, V.R. (1992a) Temperature effects on cotton fruit retention. *Agronomy Journal* 84, 26–30.

Reddy, K.R., Reddy, V.R. and Hodges, H.F. (1992b) Temperature effects on early season cotton growth and development. *Agronomy Journal* 84, 229–237.

Reddy, K.R., Hodges, H.F. and McKinion, J.M. (1995) Cotton crop responses to a changing environment. Climate change and agriculture: Analysis of potential international impacts. *Proceedings of a Symposium* sponsored by the American Society of Agronomy in Minneapolis, 4–5 November 1992. Organized by Division A-3 (Agroclimatology and Agronomic Modeling) and Division A-6 (International Agronomy), pp. 3–30.

Reddy, K.R., Hodges, H.F., McCarty, W.H. and McKinion, J.M. (1996) *Weather and Cotton Growth: Present and Future*. Office of Agricultural Communications, Division of Agriculture, Forestry, and Veterinary Medicine, Mississippi State University, Mississippi, pp. 1–23.

Reddy, K.R., Hodges, H.F. and McKinion, J.M. (1997a) A comparison of scenarios for the effect of global climate change on cotton growth and yield. *Australian Journal of Plant Physiology* 24, 707–713.

Reddy, K.R., Hodges, H.F. and McKinion, J.M. (1997b) Crop modeling and applications: A cotton example. *Advances in Agronomy* 59, 225–290.

Reddy, K.R., Hodges, H.F. and McKinion, J.M. (1997c) Modeling temperature effects on cotton internode and leaf growth. *Crop Science* 37, 503–509.

Reddy, K.R., Robana, R.R., Hodges, H.F., Liu, X.J. and McKinion, J.M. (1998) Interactions of CO_2 enrichment and temperature on cotton growth and leaf characteristics. *Environmental and Experimental Botany* 39, 117–129.

Reddy, K.R., Davidonis, G.H., Johnson, A.S. and Vinyard, B.T. (1999) Temperature regime and carbon dioxide enrichment alter cotton boll development and fiber properties. *Agronomy Journal* 91, 851–858.

Reddy, K.R., Hodges, H.F. and Kimball, B.A. (2000) Crop ecosystem responses to global climate change: Cotton. In: Reddy, K.R. and Hodges, H.F. (eds) *Climate Change and Global Crop Productivity*. CAB International, Wallingford, UK, pp. 161–187.

Reddy, K.R., Hodges, H.F., Read, J.J., McKinion, J.M., Baker, J.T., Tarpley, L. and Reddy, V.R. (2001) Soil-plant-atmosphere-research (SPAR) facility: a tool for plant research and modeling. *Biotronics* 30, 27–50.

Reddy, K.R., Kakani, V.G., McKinion, J.M. and Baker, D.N. (2002) Applications of a cotton simulation model, GOSSYM, for crop management, economic and policy decisions. In: Ahuja, L.R., Ma, L. and Howell, T.A. (eds) *Agricultural System Models in Field Research and Technology Transfer*. CRC Press, LLC, Boca Raton, Florida, pp. 33–73.

Reddy, K.R., Prasad, P.V.V. and Kakani, V.G. (2005) Crop responses to elevated carbon dioxide and interactions with temperature. In: Tuba, Z. (ed.) *Ecological Responses and Adaptations of Crops to Rising Atmospheric Carbon Dioxide*. Food Products Press, Binghamton, New York, pp. 157–191.

Reddy, K.R., Kakani, V.G. and Hodges, H.F. (2008) Exploring the use of environmental productivity index concept for crop production and modeling. In: Ahuja, L.R., Reddy, V.R. and Saseendran, S.A. (eds) *Response of Crops to Limited Water: Understanding and Modeling of Water Stress Effects on Plant Growth Processes*. ASA, CSSA, and SSSA, Madison, Wisconsin, pp. 387–410.

Reddy, V.R., Baker, D.N. and Hodges, H.F. (1991a) Temperature effects on cotton canopy growth, photosynthesis, and respiration. *Agronomy Journal* 83, 699–704.

Reddy, V.R., Reddy, K.R. and Baker, D.N. (1991b) Temperature effect on growth and development of cotton during the fruiting period. *Agronomy Journal* 83, 211–217.

Reddy, V.R., Reddy, K.R. and Hodges, H.F. (1995) Carbon dioxide enrichment and temperature effects on cotton canopy photosynthesis, transpiration, and water-use efficiency. *Field Crops Research* 41, 13–23.

Reeves, D.W. (1994) Cover crops and rotations. In: Hatfield, J.L. and Stewart, B.A. (eds) *Advances in Soil Science – Crops Residue Management*. Lewis Publishers, CRC Press, Boca Raton, Florida, pp. 125–172.

Reicosky, D.C., Meyer, W.S., Schaefer, N.L. and Sides, R.D. (1985) Cotton response to short-term waterlogging imposed with a water-table gradient facility. *Agricultural Water Management* 10, 127–143.

Reid, C.D., Maherali, H., Johnson, H.B., Smith, S.D., Wullschleger, S.D. and Jackson, R.B. (2003) On the relationship between stomatal characters and atmospheric CO_2. *Geophysical Research Letters* 30. DOI: 198310.1029/2003gl017775.

Reuveni, J. and Gale, J. (1985) The effect of high-levels of carbon-dioxide on dark respiration and growth of plants. *Plant Cell and Environment* 8, 623–628.

Richards, D., Yeates, S., Roberts, J. and Gregory, R. (2006) Does Bollgard II® cotton use more water? *13th Australian Cotton Conference*, Cotton Australia, Broadbeach, Queensland, Australia.

Richards, Q.D., Bange, M.P. and Johnston, S.B. (2008) HydroLOGIC: An irrigation management system for Australian cotton. *Agricultural Systems* 98, 40–49.

Ritchie, J.W., Abawi, G.Y., Dutta, S.C., Harris, T.R. and Bange, M. (2004) Risk management strategies using seasonal climate forecasting in irrigated cotton production: a tale of stochastic dominance. *Australian Journal of Agricultural and Resource Economics* 48, 65–93.

Roberts, G.N. and Constable, G.A. (2003) Impact of crop management on cotton crop maturity and yield. *11th Australian Agronomy Conference*, Australian Agronomy Society, Geelong, Victoria, Australia.

Rochester, I., Constable, G. and Saffigna, P. (1996) Effective nitrification inhibitors may improve fertilizer recovery in irrigated cotton. *Biology and Fertility of Soils* 23, 1–6.

Rochester, I.J. (2003) Estimating nitrous oxide emissions from flood-irrigated alkaline grey clays. *Australian Journal of Soil Research* 41, 197–206.

Rochester, I.J. (2007) Nutrient uptake and export from an Australian cotton field. *Nutrient Cycling Agroecosystems* 77, 213–223.

Rochester, I.J. (2011) Sequestering carbon in minimum-tilled clay soils used for irrigated cotton and grain production. *Soil and Tillage Research* 112, 1–7.

Rochester, I.J. and Constable, G.A. (2015) Improvements in nutrient uptake and nutrient use-efficiency in cotton cultivars released between 1973 and 2006. *Field Crops Research* 173, 14–21.

Rochester, I.J. and Peoples, M.B. (2005) Growing vetches (*Vicia villosa* Roth) in irrigated cotton systems: inputs of fixed N, N fertiliser savings and cotton productivity. *Plant and Soil* 271, 251–264.

Rochester, I.J., Peoples, M.B. and Constable, G.A. (2001a) Estimation of the fertiliser requirement of cotton grown after legume crops. *Field Crops Research* 70, 43–53.

Rochester, I.J., Peoples, M.B., Hulugalle, N.R., Gault, R.R. and Constable, G.A. (2001b) Using legumes to enhance nitrogen fertility and improve soil condition in cotton cropping systems. *Field Crop Abstracts* 70, 27–41.

Rodriguez, D., Cox, H., deVoil, P. and Power, B. (2014) A participatory whole farm modelling approach to understand impacts and increase preparedness to climate change in Australia. *Agricultural Systems* 126, 50–61.

Rosolem, C.A., Oosterhuis, D.M. and Souza, F.S.d. (2013) Cotton response to mepiquat chloride and temperature. *Scientia Agricola* 70, 82–87.

Roth, G., Harris, G., Gillies, M., Montgomery, J. and Wigginton, D. (2013) Water-use efficiency and productivity trends in Australian irrigated cotton: a review. *Crop and Pasture Science* 64, 1033–1048.

Rowland-Bamford, A.J., Baker, J.T., Allen, L.H. and Bowes, G. (1991) Acclimation of rice to changing atmospheric carbon-dioxide concentration. *Plant Cell and Environment* 14, 577–583.

Ruidisch, M., Bartsch, S., Kettering, J., Huwe, B. and Frei, S. (2013) The effect of fertilizer best management practices on nitrate leaching in a plastic mulched ridge cultivation system. *Agriculture Ecosystems & Environment* 169, 21–32.

Runion, G.B., Curl, E.A., Rogers, H.H., Backman, P.A., Rodriguezkabana, R. and Helms, B.E. (1994) Effects of free air CO_2 enrichment on microbial populations in the rhizosphere and phyllosphere of cotton. *Agricultural and Forest Meteorology* 70, 117–130.

Rustad, L.E., Campbell, J.L., Marion, G.M., Norby, R.J., Mitchell, M.J., Hartley, A.E., Cornelissen, J.H.C. and Gurevitch, J. (2001) A meta-analysis of the response of soil respiration, net nitrogen mineralization, and aboveground plant growth to experimental ecosystem warming. *Oecologia* 126, 543–562.

Sadok, W. and Sinclair, T.R. (2009) Genetic variability of transpiration response to vapor pressure deficit among soybean cultivars. *Crop Science* 49, 955–960.

Sadras, V.O. and Milroy, S.P. (1996) Soil-water thresholds for the responses of leaf expansion and gas exchange: A review. *Field Crops Research* 47, 253–266.

Sage, R.F. (1994) Acclimation of photosynthesis to increasing atmospheric CO_2 - the gas-exchange perspective. *Photosynthesis Research* 39, 351–368.

Sage, R.F. and Coleman, J.R. (2001) Effects of low atmospheric CO_2 on plants: more than a thing of the past. *Trends in Plant Sciences* 6, 18–24.

Sage, R.F., Sharkey, T.D. and Seemann, J.R. (1989) Acclimation of photosynthesis to elevated CO_2 in 5 C_3 species. *Plant Physiology* 89, 590–596.

Sahay, R.K. (1989) Photosynthetic and stomatal responses of cotton to drought stress and waterlogging. *Agricultural Science Digest* 9, 198–200.

Salvucci, M.E. and Crafts-Brandner, S.J. (2004) Inhibition of photosynthesis by heat stress: the activation state of Rubisco as a limiting factor in photosynthesis. *Physiologia Plantarum* 120, 179–186.

Samarakoon, A.B. and Gifford, R.M. (1995) Soil water content under plants at high CO_2 concentration and interactions with the direct CO_2 effects: a species comparison. Terrestrial ecosystem interactions with global change. In: *The First GCTE Science Conference*, Woods Hole, Massachusetts, 23–27 May 1994, pp. 193–202.

Samarakoon, A.B. and Gifford, R.M. (1996) Water use and growth of cotton in response to elevated CO_2 in wet and drying soil. *Australian Journal of Plant Physiology* 23, 63–74.

Sankaranarayanan, K., Praharaj, C.S., Nalayini, P., Bandyopadhyay, K.K. and Gopalakrishnan, N. (2010) Climate change and its impact on cotton (*Gossypium* sp.). *Indian Journal of Agricultural Sciences* 80, 561–575.

Saranga, Y., Jiang, C.X., Wright, R.J., Yakir, D. and Paterson, A.H. (2004) Genetic dissection of cotton physiological responses to arid conditions and their inter-relationships with productivity. *Plant Cell and Environment* 27, 263–277.

Saranga, Y., Sass, N., Tal, Y. and Yucha, R. (1998) Drought conditions induce mote formation in interspecific cotton hybrids. *Field Crops Research* 55, 225–234.

Sardans, J. and Penuelas, J. (2005) Drought decreases soil enzyme activity in a Mediterranean *Quercus ilex* L. forest. *Soil Biology & Biochemistry* 37, 455–461.

Sardans, J. and Penuelas, J. (2010) Soil enzyme activity in a Mediterranean forest after six years of drought. *Soil Science Society of America Journal* 74, 838–851.

Schellhorn, N.A., Bianchi, F.J. and Hsu, C.L. (2014) Movement of entomophagous arthropods in agricultural landscapes: links to pest suppression. *Annual Review of Entomology* 59, 559–581.

Schiermeier, Q. (2015) Quest for climate-proof farms. *Nature* 523, 396–397.

Schomberg, H.H., Fisher, D.S., Reeves, D.W., Endale, D.M., Raper, R.L., Jayaratne, K.S.U., Gamble, G.R. and Jenkins, M.B. (2014) Grazing winter rye cover crop in a cotton no-till system: yield and economics. *Agronomy Journal* 106, 1041–1050.

Setter, T.L., Waters, I., Sharma, S.K., Singh, K.N., Kulshreshtha, N., Yaduvanshi, N.P.S., Ram, P.C., Singh, B.N., Rane, J., McDonald, G., Khabaz-Saberi, H., Biddulph, T.B., Wilson, R., Barclay, I., McLean, R.

and Cakir, M. (2009) Review of wheat improvement for waterlogging tolerance in Australia and India: the importance of anaerobiosis and element toxicities associated with different soils. *Annals of Botany* 103, 221–235.

Shen, J.B., Li, C.J., Mi, G.H., Li, L., Yuan, L.X., Jiang, R.F. and Zhang, F.S. (2013) Maximizing root/rhizosphere efficiency to improve crop productivity and nutrient use efficiency in intensive agriculture of China. *Journal of Experimental Botany* 64, 1181–1192.

Shogren, R.L. (2001) Biodegradable mulches from renewable resources. *Journal of Sustainable Agriculture* 16, 33–47.

Singh, B.K., Bardgett, R.D., Smith, P. and Reay, D.S. (2010) Microorganisms and climate change: terrestrial feedbacks and mitigation options. *Nature Reviews Microbiology* 8, 779–790.

Smith, P. (2008) Land use change and soil organic carbon dynamics. *Nutrient Cycling in Agroecosystems* 81, 169–178.

Smith, P., Martino, D., Cai, Z., Gwary, D., Janzen, H., Kumar, P., McCarl, B., Ogle, S., O'Mara, F., Rice, C., Scholes, B., Sirotenko, O., Howden, M., McAllister, T., Pan, G., Romanenkov, V., Schneider, U., Towprayoon, S., Wattenbach, M. and Smith, J. (2008) Greenhouse gas mitigation in agriculture. *Philosophical Transactions of the Royal Society B: Biological Sciences* 363, 789–813.

Snider, J.L., Oosterhuis, D.M., Skulman, B.W. and Kawakami, E.M. (2009) Heat stress-induced limitations to reproductive success in *Gossypium hirsutum*. *Physiologia Plantarum* 137, 125–138.

Snider, J.L., Oosterhuis, D.M. and Kakwakami, E.M. (2010) Diurnal pollen tube growth rate is slowed by high temperature in field-grown *Gossypium hirsutum* pistils. *Journal of Plant Physiology* 168, 441–448.

Snider, J.L., Oosterhuis, D.M. and Kawakami, E.M. (2011) Mechanisms of reproductive thermotolerance in *Gossypium hirsutum*: the effect of genotype and exogenous calcium application. *Journal of Agronomy and Crop Science* 197, 228–236.

Sorenson, R.B., Butts, C.L. and Nutti, R.C. (2011) Deep subsurface drip irrigation for cotton in the southeast. *Journal of Cotton Science* 15, 233–242.

Stark, J.M. and Firestone, M.K. (1995) Mechanisms for soil moisture effects on activity of nitrifying bacteria. *Applied and Environmental Microbiology* 61, 218–221.

Stathakos, T.D., Gemtos, T.A., Tsatsarelis, C.A. and Galanopoulou, S. (2006) Evaluation of three cultivation practices for early cotton establishment and improving crop profitability. *Soil and Tillage Research* 87, 135–145.

Stidham, M.A., Uribe, E.G. and Williams, G.J.I. (1982) Temperature dependence of photosynthesis in *Agropyron smithii* Rydb. II Contribution from electron transport and photophosphorylation. *Plant Physiology* 69, 292–334.

Stiller, W.N., Reid, P.E. and Constable, G.A. (2004) Maturity and leaf shape as traits influencing cotton cultivar adaptation to dryland conditions. *Agronomy Journal* 96, 656–664.

Stiller, W.N., Read, J.J., Constable, G.A. and Reid, P.E. (2005) Selection for water use efficiency traits in a cotton breeding program: Cultivar differences. *Crop Science* 45, 1107–1113.

Stockton, J.R. and Walhood, V.T. (1960) Effect of irrigation and temperature on fiber properties. In: *14th Annual Beltwide Cotton Defoliation Physiology Conference*, National Cotton Council, Memphis, Tennessee, pp. 11–14.

Stokes, C.J. and Howden, S.M. (2010) Summary. In: Stokes, C. and Howden, M. (eds) *Adapting Agriculture to Climate Change: Preparing Australian Agriculture, Forestry and Fisheries for the Future*. CSIRO Publishing, Collingwood, Victoria, Australia, pp. 257–268.

Stone, R. and Auliciems, A. (1992) SOI phase relationships with rainfall in eastern Australia. *International Journal of Climatology* 12, 625–636.

Storch, D.K. and Oosterhuis, D.M. (2009) *Effect of 1-Methylcyclopropene on the Growth and Biochemistry of Heat-Stressed Cotton in a Controlled Environment*. University of Arkansas Agriculture Experimental Station, Research Series, pp. 58–62.

Tans, P. and Keeling, R. (2015) *Trends in Atmospheric Carbon Dioxide*. National Oceanic and Atmospheric Administration Earth System Research Laboratory Global Monitoring Division, Scripps Institution of Oceanography, University of California, San Diego, California.

Tennakoon, S.B. and Milroy, S.P. (2003) Crop water use and water use efficiency on irrigated cotton farms in Australia. *Agricultural Water Management* 61, 179–194.

Thomas, R.B. and Strain, B.R. (1991) Root restriction as a factor in photosynthetic acclimation of cotton seedlings grown in elevated carbon dioxide. *Plant Physiology (Bethesda)* 96, 627–634.

Thomas, R.B., Reid, C.D., Ybema, R. and Strain, B.R. (1993) Growth and maintenance components of leaf respiration of cotton grown in elevated carbon-dioxide partial pressure. *Plant Cell and Environment* 16, 539–546.

Thorp, K.R., Ale, S., Bange, M.P., Barnes, E.M., Hoogenboom, G., Lascano, R.J., McCarthy, A.C., Nair, S., Paz, J.O., Rajan, N., Reddy, K.R., Wall, G.W. and White, J.W. (2014) Development and application of process-based simulation models for cotton production: a review of past, present, and future directions. *Journal of Cotton Science* 18, 10–47.

Tissue, D.T. and Oechel, W.C. (1987) Response of *Eriophorum vaginatum* to elevated CO_2 and temperature in the Alaskan tussock tundra. *Ecology* 68, 401–410.

Tissue, D.T., Thomas, R.B. and Strain, B.R. (1993) Long-term effects of elevated CO_2 and nutrients on photosynthesis and rubisco in loblolly-pine seedlings. *Plant Cell and Environment* 16, 859–865.

Tissue, D.T., Griffin, K.L., Thomas, R.B. and Strain, B.R. (1995) Effects of low and elevated CO_2 on C-3 and C-4 annuals. 2. Photosynthesis and leaf biochemistry. *Oecologia* 101, 21–28.

Toulmin, C. (2009) *Climate Change in Africa*. ZED Books, London.

Travis, A.J. and Mansfield, T.A. (1979) Reversal of the CO_2-responses of stomata by fusicoccin. *The New Phytologist* 83, 607–614.

Truman, C.C., Reeves, D.W., Shaw, J.N., Motta, A.C., Burmester, C.H., Raper, R.L. and Schwab, E.B. (2003) Tillage impacts on soil property, runoff, and soil loss variations from a Rhodic Paleudult under simulated rainfall. *Journal of Soil and Water Conservation* 58, 258–267.

Turkey (2007) *First National Communication of Turkey under the United Nations Framework Convention on Climate Change, United Nations Framework Convention on Climate Change.* UNFCCC, Switzerland, 184 pp.

UNFCCC (2008) *Climate Change: Impacts, Vulnerabilities and Adaptation in Developing Countries, United Nations Framework Convention on Climate Change.* UNFCCC, Switzerland, 68 pp.

Unger, I.M., Motavalli, P.P. and Muzika, R.M. (2009) Changes in soil chemical properties with flooding: A field laboratory approach. *Agriculture Ecosystems & Environment* 131, 105–110.

Unger, P.W. and Vigil, M.F. (1998) Cover crop effects on soil water relationships. *Journal of Soil and Water Conservation* 53(3), 200–207.

Uzbekistan (2008) *Second National Communication of the Republic of Uzbekistan under the United Nations Framework Convention on Climate Change, United Nations Framework Convention on Climate Change.* UNFCCC, Switzerland, 184 pp.

van der Sluijs, M.H., Long, R.L. and Bange, M.P. (2015) Comparing cotton fiber quality from conventional and round module harvesting methods. *Textile Research Journal* 85(9), 987–997.

Walthall, C.L., Hatfield, J., Backlund, P., Lengnick, L., Marshall, E., Walsh, M., Adkins, S., Aillery, M., Ainsworth, E.A., Ammann, C., Anderson, C.J., Bartomeus, I., *et al.* (2012) *Climate Change and Agriculture in the United States: Effects and Adaptation.* USDA, Washington, DC, 186 pp.

Wang, Y., Xie, Z., Malhi, S.S., Vera, C.L., Zhang, Y. and Wang, J. (2009) Effects of rainfall harvesting and mulching technologies on water use efficiency and crop yield in the semi-arid Loess Plateau, China. *Agricultural Water Management* 96, 374–382.

Wanjura, D.F. and Barker, G.L. (1985) Cotton lint yield accumulation rate and quality development. *Field Crops Research* 10, 205–218.

Wanjura, D.F., Hudspeth, E.B.J. and Bilbro, J.D. (1969) Emergence time, seed quality, and planting depth effects on yield and survival of cotton (*Gossypium hirsutum* L.). *Agronomy Journal* 61, 63–65.

Watanabe, C.K., Sato, S., Yanagisawa, S., Uesono, Y., Terashima, I. and Noguchi, K. (2014) Effects of elevated CO_2 on levels of primary metabolites and transcripts of genes encoding respiratory enzymes and their diurnal patterns in *Arabidopsis thaliana*: Possible relationships with respiratory rates. *Plant and Cell Physiology* 55, 341–357.

Wiegand, C.L. and Namken, L.N. (1966) Influences of plant moisture stress solar radiation and air temperature on cotton leaf temperature. *Agronomy Journal* 58, 582–586.

Williams, A., White, N., Mushtaq, S., Cockfield, G., Power, B. and Kouadio, L. (2015) Quantifying the response of cotton production in eastern Australia to climate change. *Climatic Change* 129, 183–196.

Wilson, L.J., Sadras, V.O., Heimoana, S.C. and Gibb, D. (2003) How to suceed by doing nothing: Cotton compensation after simulated early season pest damage. *Crop Science* 73, 2125–2134.

Wilson, L.J., Mensah, R.K. and Fitt, G.P. (2004) Implementing integrated pest management in Australian cotton. In: Horowitz, A.R. and Ishaaya, I. (eds) *Novel Approaches to Insect Pest Management in Field and Protected Crops.* Springer-Verlag, Berlin, pp. 97–118.

Wong, S.C. (1979) Elevated atmospheric partial pressure of CO_2 and plant growth. I. Interaction of nitrogen and photosynthetic capacity in C3 and C4 plants. *Oecologia (Berlin)* 44, 68–74.

Wong, S.C. (1990) Elevated atmospheric partial pressure of CO_2 and plant growth. *Photosynthesis Research* 23, 171–180.

Wood, C.W., Torbert, H.A., Rogers, H.H., Runion, G.B. and Prior, S.A. (1994) Free-air CO_2 enrichment effects on soil carbon and nitrogen. *Agricultural and Forest Meteorology* 70, 103–116.

Wortman, S.E., Francis, C.A., Bernards, M.L., Drijber, R.A. and Lindquist, J.L. (2012) Optimizing cover crop benefits with diverse mixtures and an alternative termination method. *Agronomy Journal* 104, 1425–1435.

Wu, G., Chen, F.J. and Ge, F. (2006) Response of multiple generations of cotton bollworm *Helicoverpa armigera* Hubner, feeding on spring wheat, to elevated CO_2. *Journal of Applied Entomology* 130, 2–9.

Wu, G., Chen, F.J., Ge, F. and Sun, Y.C. (2007) Effects of elevated carbon dioxide on the growth and foliar chemistry of transgenic Bt cotton. *Journal of Integrative Plant Biology* 49, 1361–1369.

Wullschleger, S.D. and Oosterhuis, D.M. (1990) Photosynthesis of individual field-grown cotton leaves during ontogeny. *Photosynthesis Research* 23, 163–170.

Yang, C.Y., Hsu, F.C., Li, J.P., Wang, N.N. and Shih, M.C. (2011) The AP2/ERF transcription factor AtERF73/HRE1 modulates ethylene responses during hypoxia in Arabidopsis. *Plant Physiology* 156, 202–212.

Yang, Y.M., Yang, Y.H., Han, S.M., Macadam, I. and Liu, D.L. (2014) Prediction of cotton yield and water demand under climate change and future adaptation measures. *Agricultural Water Management* 144, 42–53.

Yeates, S.J., Richards, D. and Roberts, J. (2010) High insect protection of GM Bt cotton changes crop morphology and response to water compared to non Bt cotton. In: Dove, H. and Culvenor, R.A. (eds) *15th Agronomy Conference: Food Security from Sustainable Agriculture.* Australian Society of Agronomy, Lincoln, New Zealand.

Yong, J.W.H., Wong, S.C. and Farquhar, G.D. (1997) Stomatal responses to changes in vapour pressure difference between leaf and air. *Plant Cell and Environment* 20, 1213–1216.

Yoon, S.T., Hoogenboom, G., Flitcroft, I. and Bannayan, M. (2009) Growth and development of cotton (*Gossypium hirsutum* L.) in response to CO_2 enrichment under two different temperature regimes. *Environmental and Experimental Botany* 67, 178–187.

Zhang, S., Sadras, V., Chen, X. and Zhang, F. (2013) Water use efficiency of dryland wheat in the Loess Plateau in response to soil and crop management. *Field Crops Research* 151, 9–18.

Zhang, W., Parker, K.M., Luo, Y., Wan, S., Wallace, L.L. and Hu, S. (2005) Soil microbial responses to experimental warming and clipping in a tallgrass prairie. *Global Change Biology* 11, 266–277.

Zhao, D. and Oosterhuis, D. (1997) Physiological response of growth chamber-grown cotton plants to the plant growth regulator PGR-IV under water-deficit stress. *Environmental and Experimental Botany* 38, 7–14.

Zhao, D.L., Reddy, K.R., Kakani, V.G., Mohammed, A.R., Read, J.J. and Gao, W. (2004) Leaf and canopy photosynthetic characteristics of cotton (*Gossypium hirsutum*) under elevated CO_2 concentration and UV-B radiation. *Journal of Plant Physiology* 161, 581–590.

Zhao D., Reddy, K.R., Kakani, V.G., Koti, S. and Gao, W. (2005) Physiological causes of cotton fruit abscission under conditions of high temperature and enhanced ultraviolet-B radiation. *Physiologia Plantarum* 124, 189–199.

Zhou, L.-M., Jin, S.-L., Liu, C.-A., Xiong, Y.-C., Si, J.-T., Li, X.-G., Gan,Y.-T. and Li, F.-M. (2012) Ridge-furrow and plastic-mulching tillage enhances maize–soil interactions: Opportunities and challenges in a semiarid agroecosystem. *Field Crops Research* 126, 181–188.

Ziska, L.H., Bunce, J.A. and Caulfield, F. (1998) Intraspecific variation in seed yield of soybean (*Glycine max*) in response to increased atmospheric carbon dioxide. *Australian Journal of Plant Physiology* 25, 801–807.

Ziska, L.H., Teasdale, J.R. and Bunce, J.A. (1999) Future atmospheric carbon dioxide may increase tolerance to glyphosate. *Weed Science* 47, 608–615.

Ziska, L.H., Bunce, J.A. and Caulfield, F.A. (2001) Rising atmospheric carbon dioxide and seed yield of soybean genotypes. *Crop Science* 41, 385–391.

Ziska, L.H., Bunce, J.A., Shimono, H., Gealy, D.R., Baker, J.T., Newton, P.C.D., Reynolds, M.P., Jagadish, K.S.V., Zhu, C., Howden, M. and Wilson, L.T. (2012) Food security and climate change: on the potential to adapt global crop production by active selection to rising atmospheric carbon dioxide. *Proceedings of the Royal Society B, Biological Sciences* 279, 4097–4105.

Zogg, G.P., Zak, D.R., Ringelberg, D.B., MacDonald, N.W., Pregitzer, K.S. and White, D.C. (1997) Compositional and functional shifts in microbial communities due to soil warming. *Soil Science Society of America Journal* 61, 475–481.